PLANT ECOLOGY

INTRODUCTION TO PLANT ECOLOGY

BY

MAURICE ASHBY

Ph.D. (London), B.Sc., A.R.C.S.
Portsmouth College of Education

MACMILLAN

ST MARTIN'S PRESS

© Maurice Ashby 1961

First edition 1961
Reprinted 1963
Reprinted with corrections 1965
Second edition 1969
Reprinted 1971

Published by
THE MACMILLAN PRESS LTD
London and Basingstoke
Associated companies in New York Toronto
Dublin Melbourne Johannesburg and Madras

SBN 333 06260 4

Printed in Great Britain by
ROBERT MACLEHOSE AND CO LTD
The University Press, Glasgow

69 - 17458

PREFACE

The study of plant communities in the field provides an opportunity which can make ecology a most stimulating ingredient of the elementary biology course. Yet there is a danger that this opportunity may be wasted. Studies of plant communities undertaken without sufficient background knowledge can prove shallow and unsatisfying to students, seemingly mere exercises in mapping, or the collection of data from sample quadrats for use in mysterious statistical rites having no very clear objective. But field observations directed simply towards recognizing some of the problems that exist within a plant community can indeed be successful and stimulating, even at an early stage in the course. Basically such problems are all concerned with the behaviour — the success or failure — of individual plants in varying local surroundings, and it is with these that the study of ecology really begins. Any true understanding of ecology must surely rest upon an appreciation of these problems and how they are met. Growth form, seasonal behaviour, depth of rooting, soil preferences, vegetative spread, pollination, genetics, indeed almost any aspect of the living plant's behaviour — all these can be viewed in terms of how they affect the plant's performance in the field and influence the community in which it lives. Ecology then becomes more than just a branch of biology; it becomes a way of thinking which colours the whole of the biology course.

This kind of outlook can be present in biology teaching from the start. It will win the enthusiasm of the young naturalist, which can later be moulded into the more scientific approach of ecology, without any need for a parting of the ways between natural history as a hobby and the more serious academic study of botany. No claim is made that there is anything particularly new about these ideas, but I have tried in this book to develop them somewhat with the object of helping at the later level. The text is not geared to any particular syllabus but has been written with sixth form, scholarship and training college students in mind. I hope that it may serve a useful purpose also in the elementary stages of some university courses. The general grounding assumed is roughly equivalent to the standard at the end of the first year in the sixth form.

Emphasis is laid on the principles involved in the ecology of individual species, and the many simple yet fascinating practical problems arising from this kind of study. In the hope that some readers may wish to try this kind of 'research' for themselves I have given details of a number of experimental techniques in the appendix. Soils, as the environment of the roots and underground storage organs with their resting buds, receive rather detailed treatment because of the rapid advances which have been made in this field of knowledge and the tendency to dismiss it rather briefly in elementary text-books. In the final section on plant communities I have made no attempt to reduplicate the excellent descriptions of British vegetation which are already available. The aim is rather to try and explain the principles at work in the formation of plant communities, how stability is maintained or changes come, and how one can set about studying them in the field. It is hoped that the reader will find sufficient guidance to enable him to build up an understanding of the local vegetation from his own studies.

Statistical analysis is playing an ever growing part in the study of plant communities, and, while most of the techniques are beyond the scope of this book, I have attempted to give the reader some idea of the way in which the methods are used. A few simple techniques are explained in detail and references given for those who wish to follow up this study.

I am very conscious of my debt to the many ecologists whose work and ideas I have drawn upon in preparing the book, and I should like to record my gratitude to them here. Also to those people who, as former pupils, have helped in the discussion of ideas and in tackling some of the experimental work described. Wherever material has been used directly the source is acknowledged in the text, and I trust that there are no omissions.

Especial thanks are due to those from whom I have received more direct help. To Mr. L. J. F. Brimble who has acted as general editor; to Mr. W. H. Dowdeswell for his unfailing encouragement and advice on many points, and for his most valuable help in supplying the photographs for Figs. 4.2a and b; 4.3 4.4; 6.2; 10.3a and b, and for preparing prints where my own negatives could be used; also to Mr. P. Greig-Smith for his generous help and advice on the quantative aspects of the study of plant communities. I should like to express my thanks also to Dr. F. Yates for his advice over the statistical material; to Dr. A. J. Rutter for advice on soil tensiometers; to Mr. Martin George for his kindness in lending the photographs for Figs. 6.4 and 6.5; and to Mr. Ben Knutson for permission to reproduce Fig. 8.5. A number of authors and publishers have kindly allowed me to reproduce diagrams and tables; the sources of these are acknowledged in the text. Finally I should like to take this opportunity of thanking my daughter Lyndall for her valuable help in preparing the index.

Maurice Ashby

Southsea, December 1960

PREFACE TO THE SECOND EDITION

The need to 'build ecological understanding into the wisdom of the race' takes on fresh urgency with the ever growing demands made by man upon his environment. In face of this, the contribution from elementary plant ecology may at first seem somewhat remote and academic. Production ecology, which has made such rapid progress during the past few years, gives an added perspective which makes the study of both plant and animal ecology more meaningful. I am therefore especially glad to have the opportunity of adding a chapter on production ecology to this second edition. Necessarily it draws widely upon the work and thinking of Professor E. P. Odum, who has contributed so much to this field, and I am happy to acknowledge my debt to him. There have also been a number of smaller additions, bringing the text and bibliography more up-to-date, and two further practical techniques are described in the appendix. Finally, some of the original photographs have been replaced.

Maurice Ashby

To
the many young
biology students who have
unwittingly contributed so
much to this book

CONTENTS

INTRODUCTION

PART I   TOLERANCE

PART II   AGGRESSION

PART III   PLANT COMMUNITIES

INTRODUCTION

I

Scope of Ecology

No living creature, plant or animal, can exist in complete isolation. If an animal, it is bound to depend upon other living creatures, ultimately plants, for its food supply; it must also depend upon the activities of plants for a continued oxygen supply for its respiration. Apart from these two basic relationships it may be affected, directly or indirectly, in countless different ways by other plants and animals living around it: other animals may prey on it or compete with it for the same food; plants may provide shelter, concealment or nesting material, and so on. In like manner, the animal will produce its own effects on the surrounding plants and animals: some it may eat or destroy, thus swaying the balance between competing species; for others it will provide food; by its burrowing or trampling activities, and through its contribution of manure it may influence the texture and fertility of the soil.

This dependence on other living things is not confined to animals. Though plants manufacture their own food by photosynthesis, they are dependant on animal respiration for at least a part of the carbon dioxide which they use as raw material. Supplies of mineral salts which they use to build up their substance can only be maintained through the activities of fungi and bacteria breaking down the organic matter left in the soil by other living creatures. Again, many plants are entirely dependant on animals for pollination or for the dispersal of their seeds. Moreover, despite the apparently peaceful relationships in plant communities, there is intense competition going on for water, nutrient salts, and above all for light.

We see, then, that *other* plants and animals, through their effects both direct and indirect, form an integral part of the environment of every living organism. In a well-defined community, such as exists in a beechwood, a water meadow or a pond, the populations of plants and animals are influenced not only by physical factors like light, temperature or humidity, but also by the complex tangle of interrelationships between the living creatures themselves. As a result, the populations of different competing species exist in a state of delicate balance, easily swayed by the slightest change in any factor.

Ecology (literally *the study of plants and animals in their homes*) seeks to explain these interrelationships between all the different members of a community, and to build up an understanding of the community as a whole. To the ecologist the reactions and behaviour of any plant or animal are like a piece of a jig-saw puzzle: he must find how it fits into the picture of the whole community. Man is seen in healthy perspective as just another piece in this grand jig-saw, and his activities in terms of the effects, good or bad, that they are likely to produce on the communities and soils from which he derives his food. Ecology thus becomes more than a study; it is a way of looking at life, always attempting an integration that may help in understanding the whole community. It draws on the most divergent branches of knowledge: physiology, soil science, climatology, history, genetics, animal behaviour, and many others all make their contribution.

The whole complex of the plants and animals forming a community, together with all the interacting physical factors of the environment really form a single unit, which has been called the **ecosystem**. This takes into account *all* the living creatures in the community, from the fungi, bacteria and nematode worms living in the soil to the mosses, caterpillars and birds up in the trees; and *all* the factors of the environment, from the composition of the soil atmosphere and soil solution to wind, length of day, relative humidity, atmospheric pollution, etc. All will have some effect on the balance of the whole. It will be seen that the final aim of ecology — the complete understanding of ecosystems — is an ideal one can scarcely hope to attain. It is nevertheless an ideal well worth pursuing, and valuable progress has been made towards it. The first stage is to simplify the problem by studying different aspects of the ecosystem independently. To begin with, one can study the **habitat**, or sum total of the conditions of the environment, apart from the organisms under consideration. The second step in breaking down the problem is to study the plants and animals of the community separately. Obviously, this distinction is very artificial and in many ways undesirable, but the methods of studying animal communities and plant communities differ so widely that their separation is justified on grounds of convenience. Plants at least are not directly dependent on animals for their food, so that a study of the ecology of plant communities alone can be fairly complete in itself. The animals of course are not ignored: full account is taken of their effects upon the plants, but their relationships with each other are not discussed. Our more limited objective in this book, then, is to gain some understanding of plant communities, and the factors which influence the balance of the various plant populations within them.

The intense competition between the different members of a plant community is at first difficult to credit. As the plants cannot destroy each other by violence, it must operate entirely through modification of the physical and chemical factors of the local environment. The larger plants help to determine the local climate (or **microclimate**) beneath them particularly if they are growing in a close stand. Factors such as light intensity, soil water, and many others are likely to be influenced in such a way as to restrict the range of other species that can survive. The larger plants which are imposing these limitations through the microclimate which they create beneath them, are described as the dominant species in the community. Often the microclimate which they create does not favour the survival of their own seedlings, and they are eventually replaced by other dominants which change the nature of the whole community. Vegetation is never quite static: there are always some changes going on.

It will be seen that we shall never make much progress in our understanding of a plant community without first knowing something about the behaviour of the *individual* species present. We need to know such things as how they react to changes in microclimate, how they tolerate shade, grazing, soil acidity, frost or drought; how they spread; and what changes in microclimate they themselves produce. This detailed study of the biology of individual species is known as **autecology**, in contrast to **syneco-**

logy, the study of the actual communities. Autecology is sometimes described as not being ecology at all, but a hotchpotch of plant physiology, genetics, floral biology and various other ingredients. It is true that all these make their contribution; but the information which they contribute is integrated to give as complete a picture as possible of the behaviour of plants *in the field*. Here lies the key to any real depth of understanding of plant communities, and synecological studies undertaken without due recognition of this are apt to yield but shallow returns. Obviously this does not mean that *no* studies of plant communities can be undertaken until the autecology of all the species present has been worked out: such a policy would mean that the synecology would be postponed indefinitely. But in his introduction to the study of ecology it is essential that the student should first gain a working knowledge of how individual plants influence their neighbours and react to different conditions in the field, that is, that he should understand the principles of autecology. He has then the equipment to interpret the relationships he finds in natural plant communities, and can better see what further detailed studies of individual species may be needed. For this reason the greater part of this book is devoted to autecology. Details of the autecology of individual species are gradually being published by the British Ecological Society as the 'Biological Flora of the British Isles'.

So far we have been dealing with very wide generalities; let us now come down to something more tangible and consider the vegetation that we would see on a *summer* train journey. It makes little difference where we go: London to Folkestone will give us as wide a variety as anywhere. As the train gathers momentum, pounding up the steady slope from the Thames basin we get our first glimpse of greenery on emerging from Elmstead Woods tunnel — the open birchwoods of Chislehurst and Petts Wood, with scattered pines and oaks. Beneath the trees is a vigorous growth of bracken with here and there a dash of colour where a patch of rosebay willowherb is coming into bloom. Little of this land has been taken for farming, and even London's ever-growing appetite for suburban building land has not entirely robbed it of its characteristic vegetation. With our next check in speed we are passing through sheer cuttings in chalk rock before plunging under the ridge of the North Downs around Knockholt. Bursting from the tunnel we get a glimpse of the sweeping downs at Poll Hill; grassland with scrub of wild roses (Fig. 9.2, p. 169), quite unlike what we have seen before, and the hills capped with dense beech-woods, like sauce poured on to a pudding. The sand- and gravel-pits before Sevenoaks tell us that we are off the chalk, and on emerging from Sevenoaks tunnel we are back again in oak and birch country, though much of the land is under cultivation. The meadows bordering the alder-lined banks of the Medway as we enter Tonbridge are again grassland of a different appearance from what we have seen previously. Leaving Tonbridge, our route lies along the heart of the Weald, where the rich soils have been so long used for agriculture that no trace is left of the natural vegetation: orchards and hop-gardens are the predominating feature. But shortly before Ashford there is again an abrupt change to birch and pine with some heather, for a few miles. For the last stage of the journey the ridge of the chalk downs edges ever closer, ob-scured for a short time by the woods around Sandling, with their sprinkling of rhododendrons.

A journey almost anywhere in England would show variety of this kind. The underlying cause can scarcely be climatic differ-ences; it is fairly obviously connected with variations in the soil. Reference to a geological map bears this out, for the changes in vegetation correspond to differences in the underlying rock. We passed through London gravels at Chislehurst, chalk at Poll Hill, greensand beyond Sevenoaks and Wealden clay after Tonbridge. Even the small area of birch-pine heath before Ashford cor-responds to a projecting tongue of greensand which we crossed, and the soils of the Sandling woodlands are derived from the same rock. If we could return to make a closer inspection of the woods and meadows seen from the train, we should find much more than mere collections of birches, beech or expanses of grass. Each woodland is a well-defined plant community, dominated by one or two species of tree, and with its own characteristic her-baceous species growing on the ground. The chalk grassland of Poll Hill with its rich variety of small herbs is entirely distinct from the river meadows outside Tonbridge. How did these differ-ent communities arise, and to what do they owe their distinctive composition and structure? The study of plant ecology is con-cerned with finding the answers to such questions as these.

The examples we have been discussing emphasize the importance of soil in determining plant communities, but the environment above the ground plays just as big a part. This is less
apparent over small areas because the climate does not vary much:
it is perhaps best seen when steep slopes with northerly and
southerly aspects are compared (Fig. 1.1). When considering only
the plant communities, as we are doing, we tend to regard the
effects of animals as biotic factors that are part and parcel of the

FIG. 1.1. Colonization of more sheltered side of the valley by hawthorn
and mountain ash dispersed by birds. On the more exposed north-
facing slope the succession is held up at the grass stage. Yorkshire
moors at about 800 ft.

habitat. Regarded in this way, grazing is a habitat factor of cardinal
importance. Not only does it have a selective effect on the ground
vegetation, but by eliminating tree seedlings it also keeps the
spread of woodlands in check and may even destroy them by
preventing regeneration. But the distinctive nature of plant communities is not a product of the habitat alone. It results from an
interplay between the habitat factors and the plants themselves,
in much the same way that any individual plant or animal is the
product of interaction between its inherited 'nature' and the
environment in which it has developed. There is at first a large
element of *chance* in what species colonize the ground. Seed dispersal is usually more or less at random, and to a large extent

chance determines which seeds get there first. What seed parents are available is also partly a matter of chance. For example, rice-grass (*Spartina townsendii*) appeared as a new species in Southampton Water some time around 1870 as a result of chance hybridization between the native cord-grass (*Spartina maritima*) and another species, *S. alterniflora*, introduced from North America during the early nineteenth century. It proved a highly successful plant in the coastal mud-flats, spreading rapidly and ousting other competitors, and it has since been introduced into a number of other localities both here and abroad as a mud-binding grass. But whether it appears as a natural colonist in any particular mud-flat must clearly depend on the chance that colonizing material happens to be growing already in the neighbourhood.

As further colonists arrive, whatever the habitat, the part played by chance diminishes, and the development of the plant community is influenced more and more by the behaviour of the plants actually present. Two aspects of this behaviour stand out: **tolerance** of the conditions of the habitat, and **aggression**. Their importance in the development of plant communities is so great that they have been taken as a basis for the first two parts of this book.

TOLERANCE

Of the various species that arrive as colonists in any area some will thrive, but for others the conditions of the habitat will be unfavourable, and they will at best make only poor growth. Probably these weaklings will soon be ousted in competition unless the habitat changes to suit them better. The conditions of the habitat have selected from the colonizing material, which arrived largely by chance. Survival depends upon a capacity to tolerate the conditions of the habitat. But, as we shall see later, the conditions of the habitat become changed by the plants themselves, through soil improvement, shading, shelter, etc. The balance of competition is thus swayed and other species come in. Eventually larger plants virtually take control of the habitat beneath them, limiting the subordinate vegetation to species which can tolerate the microclimate they impose. This gives the community a definite structure. There are many different microclimates within the community as a whole; each offers an **ecological**

niche — a way of making a living — to species that can tolerate the local conditions.

We see, then, that it is of vital importance to understand something of the nature of this tolerance. Some species have a wide range of tolerance and are found in widely differing habitats. Others are specially adapted to tolerate extreme conditions of one kind or another. These adaptations may be physiological or structural: they may arise as a direct reaction to the conditions of the habitat (for example, manner of growth), or they may be genetically distinct forms of variable species which have been selected by the rigours of the habitat. In Part I we shall therefore be particularly concerned with the way in which different factors of the environment affect the individual plant, and how the plant reacts to them.

AGGRESSION*

A plant community emerges as a result of successful competition by the species which can tolerate and exploit the microhabitats it offers. To succeed in competition a species must not only hold its ground and maintain its numbers, but also spread to new ground at the expense of other species. The principal means of achieving this is almost invariably the shading out of rivals. We are therefore concerned with two aspects of the plant's biology when studying aggression: (1) the shade it casts and (2) its manner of spreading.

The shading effect is rather more difficult to assess than might at first appear. Not only does it depend upon the intensity of shading and the height of the leaves casting the shade, but also the seasonal variation is vitally important. For this we must know the time when the plant comes into leaf in spring, and when it dies back, or the leaves are shed. Rate of decay of the dead material left in the autumn may also be significant. Bracken comes into leaf late in the spring, but the dead fronds are so resistant to decay that they exert a continuous shading effect all the year round. Recent work on woodland communities suggests that sunflecks due to small gaps in the tree canopy, and the spectral quality of the light getting through are both of some importance to the ground vegetation.

But shading alone can scarcely constitute effective aggression;

* The term 'interference' has been suggested by Harper (1961).

it must be accompanied by spread of territory. Both vegetative spread and seed dispersal must be considered, with all the factors that influence them. Nor can we stop at seed dispersal; the cycle must be complete; the seed germinated and the new plant established before aggression is possible. Furthermore, pollination is normally an essential preliminary to seed production, and the factors affecting it have therefore a bearing on aggression. But indirectly pollination plays a much more important part in aggression taken in its widest sense. As the mechanism by which cross-fertilization and gene exchange are brought about it provides the variation needed for evolution — the evolution of more vigorous types, or varieties adapted to live in specialized habitats. In the long run, this makes the biggest contribution of all to a plant's success. A good example is provided by the rice-grass already mentioned (p. 7), for it replaces the original cord-grass wherever the two are in competition.

APPLIED ECOLOGY

This brief introduction would not be complete without some mention of the importance of applied ecology in human affairs. Primitive man was a scarce animal living in harmony with the ecosystem in which he found himself. But as he has grown in technical skill, and his numbers have increased he has played an ever greater part in upsetting the balance of nature. In countries with relatively small populations this rape of natural resources has proceeded with little heed for the future. Only in the last fifty years has the realization gradually come that conservation of natural resources is a matter of urgency if the world's growing population is to be fed. Conservation may involve specialized work in widely divergent fields; agriculture, forestry, drainage, or engineering projects such as dam building and irrigation, to name a few. But the planning of conservation policy, and the integration of all these varied activities is essentially a problem of ecology. Without an ecological approach the narrow views of specialization may be blind to blunders which stultify costly efforts for improvement. Oosting (1956) quotes cases in the United States where dams built for water conservation have become virtually useless within 10–15 years through silting up of the reservoirs. There is no fault in the dam construction, but with no control of land management in the catchment areas, lumbering and excessive grazing have continued, fostering rapid soil erosion. Where the plans for the whole operation have been co-ordinated and the ecological point of view has prevailed, as in the classic case of the Tennessee Valley Scheme, the effect on the resources of the area has been tremendous — and lasting. But schemes such as this are aimed at remedying the catastrophic effects of soil erosion resulting from the mismanagement of land in the past. As one American writer (Leopold, 1949) has put it: 'Practices we now call conservation are to a large extent, local alleviations of biotic pain. They are necessary, but they must not be confused with cures. The art of land doctoring is being pursued with vigour, but the science of land health is yet to be born.' It may fairly be said that in the twelve years since this was written, the birth of land health as a science has taken place — in the ecological work of university departments and such official organizations as the Nature Conservancy — but the infant is scarcely recognized by the public yet. Here we lag behind United States, where much more has been done to stimulate a public awareness of the importance of conservation.

A positive policy of land health would entail planned utilization of land according to its productive capacity, with management that would ensure *maintained* productivity. By 'productivity' we do not necessarily refer to food crops; it may be grazing, forest, or land set aside for æsthetic or other purposes like a catchment area or a nature reserve. Wherever it is, a given amount of solar energy falls on every acre of land in the year, depending upon latitude, amount of cloud, etc. If there is no vegetation present, it will only serve to warm and dry out the soil before it is re-radiated and lost as entropy. The productivity of the land will depend on what plant community the soil and climate can support to fix solar energy in organic matter. This fixed energy will pass from plant to animal, and from one animal to another up the food chain, until it is eventually dispersed in respiration. Thus the land will 'produce' and carry its animal community, but this is entirely limited by the amount of energy fixed by the plant community in the first place. This, in turn depends (apart from climate, which is not expendable) on two priceless assets: the soil and water. It is these we must conserve to *maintain* productivity.

Soils are gradually built up over hundreds or even thousands of years, largely through the activities of the plants themselves, but also much influenced by climate. Under natural conditions they are usually clothed with the type of vegetation they can best support — deciduous woodland in Great Britain, coniferous forests further north, grassland in low-rainfall regions like the American prairies or the steppes of Central Asia. But the natural vegetation is rarely the most suitable for man, and he modifies it to suit his own needs, thus upsetting the balance of nature. Where this is done on a large scale, without appreciation of the ecological principles involved, there is likely to be a decline in productivity through deterioration of the soil and its capacity for storing water. In harsher climates than our own there may be wholesale erosion of the topsoil following lumbering or overgrazing. A low rainfall may often be enough to give moderate productivity, provided the vegetation cover is sufficient to hold the infrequent rains and allow the soil to store up some of the water. To disturb the balance of nature in such regions means, almost certainly, that the desert will encroach on them. The impressive ruins of Roman cities along the now barren North African coast are a witness to the productivity of the land within historic times. In our own more genial climate, the deterioration is not so rapid, and our ancestors were able to learn by gradually accumulated experience how to maintain fertility in the lowland soils. The fervour which the true farmer devotes to husbanding the resources of his soil reflects the struggle that must have gone to the accumulation of this knowledge. But in upland Britain we have deterioration on a big scale, for much of the present moorland was once clothed with woods. Here the topsoil has not so much been washed away bodily, but the nutrient elements, particularly calcium, have been leached out. Much attention has been devoted to the possibility of raising the productivity of these marginal lands as rough pastures and forests. This is essentially an ecological problem; and indeed, ley farming and many other advances in pasture management owe much to our increased knowledge of the ecology of grassland.

With industrialization the demand for water for other purposes than direct plant or animal requirements is forever growing. It has recently been stated (Ovington, 1958) that it takes just over 3,000 gallons of water to make a ton of steel; about 2,500 gallons are needed to manufacture an average car, and 1 gallon to produce a pint of beer. Ecology is again of importance in problems of land management in the catchment areas which supply the water reservoirs for big cities. There is evidence that in some areas the water-table is lower than it used to be, indicating a fall in the quantities of water stored in the soil.

Quite apart from questions of conservation, agriculture has brought other ecological problems in its train, especially in the way in which it changes the balance of nature among pests and fungal parasites. In untouched communities these tend to be kept in check by natural predators, or merely by the fact that the host plants are widely scattered. But monoculture, the cultivation of pure stands of single species, by doing away with variety brings unrivalled opportunities for plagues and epidemics, with inevitable losses or the heavy expenses of keeping the damage within bounds. The ecological solution is biological control, which may sometimes involve no more than leaving intact other vegetation, such as hedgerows, which provides the natural habitats for animals preying on the pests.

2

Environment of the Roots and their Functions

Because of the difficulty of unearthing roots intact we tend to overlook the extent of the root systems even of small plants, while the ramifications and depth of tree roots must remain largely a matter of guess-work. Sometimes when gardening one gets a hint of their spread on coming across stout roots of a tree well outside the canopy formed by the foliage. But it is only by patient excavation, and washing away the soil from around them with a gentle stream of water that one can fully appreciate how richly branched the roots are. Priestley and Scott (1955) quote a case in which the entire root system of a single wild oat plant, grown free from competition for 80 days, was laid bare in this way: the total length of the root branches was estimated at fifty-four miles. It is a sobering thought to consider what a complex tangle the competing root systems under a grass meadow must form. Small wonder that root parasites like eyebright (*Euphrasia*) and yellow-rattle (*Rhinanthus*) have evolved which exploit the opportunities presented by the tangle of root branches in such situations.

The soil forms the environment in which this vast ramification is built up and carries on its activities, and it is from the soil that the green plant derives its water supply, together with all the elements of which it is composed, except carbon and oxygen. Clearly, it is of the utmost importance to understand how this complex environment can vary, and how the variations affect the activities of plant roots and other underground parts, such as rhizomes, corms and bulbs. Our understanding of these things is as yet far from complete, but it is a field in which there has been rapid progress in the last few decades, and where there is much scope for original observation.

A soil, if it is to support rich vegetation, must provide a medium in which plant roots can grow and obtain anchorage, at the same time taking up steady supplies of water and the mineral salts they require. Anchorage of herbaceous or shrubby plants normally presents few problems, and indeed trees are rarely excluded from growing on any particular type of soil for reasons of poor anchorage alone. Healthy root growth and absorption are greatly influenced by the amount of oxygen present in the soil, and, of course, there must be adequate supplies of water and mineral salts available for absorption. We see, then, that it is the *spaces* between the rock particles of the soil which are of most *immediate importance* in determining its nature, for in them are held the soil solution and the soil atmosphere upon which the well-being of the plant roots depend. But the interactions of different factors affecting the composition of soil solution and soil atmosphere are extremely complex, and, above all, soils, like vegetation, are dynamic systems going through a series of gradual changes in development. We shall therefore be in a better position to understand them if we first consider briefly the way in which soils are formed.

The first stages in the *formation of soil* from exposed rock are those of mechanical disintegration by weathering, and here the cracking and fragmenting action of frost on the rain-soaked rock masses plays a major part. In mountainous areas this process is greatly accelerated, leading to the accumulation of rock fragments in unstable 'embryo soils' and screes. Obviously, the rate of breakdown will vary greatly with the nature of the rock material, being slowest on hard, igneous rocks like granite and basalt. Where there are good cleavage planes as in slaty rocks, the roots of colonizing plants may play an important part in helping to force open the fissures. This is very noticeable, for example, at Morte Point in North Devon, where plants such as buck's-horn plantain (*Plantago coronopus*), rock sea spurrey (*Spergularia rupicola*) and especially thrift (*Armeria maritima*) can be found with stout, woody roots penetrating far into the rocks along the cleavage planes (Figs. 2.1 and 2.2). A well-established plant of thrift may have roots penetrating at least three or four feet into the rock, and it is no easy task to excavate these root systems. Even a tiny rosette of buck's-

FIG. 2.1. Colonization of slaty
rock by thrift (*Armeria maritina*).
Morte Point, North Devon

FIG. 2.2. Colonization of schistose rocks at Start Point, Devon, by
lichens, sea plantain (*P. maritima*), thrift and samphire (*Crithmum
maritimum*)

horn plantain, scarcely half an inch across, was found to have roots extending nearly six inches into the rock (Figs. 2.3 and 2.4). On a larger scale, it has been shown in a study of the limestone crags to the east of Sheffield (Jackson & Sheldon, 1949) how important a part yew trees play in breaking up the edges of the cliffs by the expanding force of their roots growing in the natural

FIG. 2.3. Soil over slate. Note penetration of grass roots

cracks of the rock (Fig. 2.5). Such processes of mechanical weathering, helped by the plants themselves, will in time yield fine rock particles, some even of colloidal dimensions, which form the raw material of soil. This mineral basis, however, is still of little value for plant growth without the addition of the soluble components of the soil solution, which result from two processes: (1) the *decay of organic matter*, largely derived from plants, and (2) the *chemical breakdown of fine rock particles*, or *chemical weathering*.

The apparent paradox that plants are helping to break down rock and contributing organic matter for decay, before a soil has

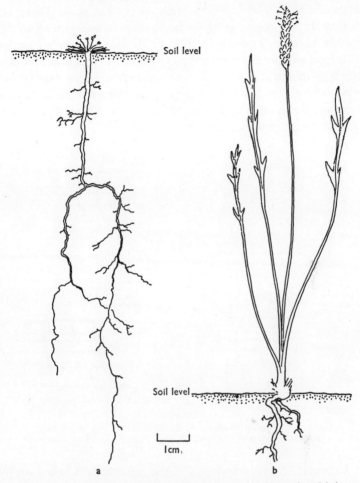

FIG. 2.4. Buck's-horn plantain (*Plantago coronopus*) growing (*a*) in cracked rock and (*b*) in soil only a few feet away

developed fit to support them, is explained by the varying requirements of different plant species, and by the uneven pace and localized nature of soil formation. Lichens are so hardy that they can flourish on bare rock surfaces, obtaining their modest mineral requirements by the direct solvent action of substances excreted by the fungal member of the partnership. In any small crevice, plant remains carried by drainage water can soon accumulate to provide, together with the mineral matter already there, the beginnings of a soil pocket in which hardy species can grow. These, on their death and decay, yield the beginnings of humus, and so the process becomes cumulative, though in detail its course varies with local conditions. An interesting parallel to this is seen in the building up of soil in the choked rain-gutters of a roof. Once a soil is established, the roots of the larger plants which it supports may penetrate down into the parent rock and help to break it up, as in the case of the yew trees mentioned above.

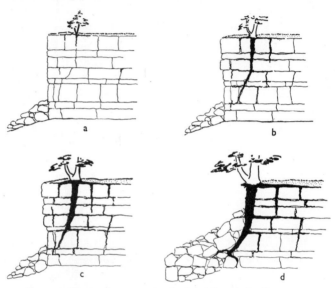

FIG. 2.5. Successive stages in the breakdown of a limestone crag by the roots of yew trees. (After Jacks and Sheldon, 1949)

Although a large proportion of the mineral salts taken up by plant roots is derived from the decay of pre-existing organic matter, we must remember that it has originated from the chemical breakdown of rocks. The pace of this process must be very gradual indeed, but it is nevertheless of prime importance as a means by which the stocks of mineral salts in circulation are being constantly replenished. As against this there is the loss from circulation of soluble mineral matter leached out of the soil by percolating water, to be carried away with drainage or deposited in lower layers of the soil as hardpan. The chemical breakdown of rock substance is achieved largely through the agency of water containing dissolved atmospheric carbon dioxide. Again the plants make their contribution in supplementing this carbon dioxide with further supplies derived from the respiration of their roots. Indeed, in soft limestones plant roots can penetrate like fungal hyphae, 'digesting' channels for themselves, often independent of any natural fissures, by means of the carbonic acid formed from excreted carbon dioxide which converts the rock into soluble calcium bicarbonate. Other acids may also play a part, notably those derived from the decay of dead plant material, and the process is further hastened by the greatly increased surface presented by the more finely divided rock particles resulting from mechanical weathering.

For our purposes it is only necessary to discuss the *chemistry of the weathering process* in general outline. The fate of the limestone, which consists of more or less pure calcium carbonate has already been mentioned. The resulting soils, though apt to be highly calcareous, are usually fertile, containing adequate supplies of other necessary elements derived from impurities in the rock and the contributions from organic remains. The igneous rocks may be regarded, broadly speaking, as consisting of complex compounds of silica and the oxides of various metals, those of aluminium, potassium, calcium, magnesium and iron being among the commonest. Thus orthoclase felspar, a regular constituent of granites, has the formula: $K_2O . Al_2O_3 . 6SiO_2$. The chief action of water is one of hydrolysis, in which the metals in the mineral complex are replaced by hydrogen and brought into solution as hydroxides; calcium, potassium and magnesium yield *strong bases*, while iron and aluminium yield *weak bases*. It is the strong bases that are dissolved out of the mineral complex most readily; aluminium oxide is particularly resistant to chemical weathering. The reaction for orthoclase felspar may be shown empirically as:

$$K_2Al_2Si_6O_{16} + 2H_2O \rightarrow 2KOH + H_2Al_2Si_6O_{16}$$
Orthoclase felspar   *Strong base*   *Remaining*
           (dissolved out)   *mineral complex'*

The soluble potassium hydroxide normally combines with acids produced in the decay of organic matter, for example, nitric acid resulting from protein breakdown, to give soluble salts. In a similar way, the remaining mineral complex may be further hydrolysed, setting free some of the silica, though this takes place less readily than the removal of the strong bases. It will be seen that with the removal of the metals and their replacement by hydrogen, the remaining mineral complex with its silica will take on a more acid character. Eventually the mineral complex may consist largely of residual silica (sand particles may be up to 95 per cent silica) or combined alumina and silica (in clays these are often present in approximately equal proportions).

Of course, soils are not necessarily formed immediately over-laying the parent rock from which they are derived; the raw mineral particles may have been transported for long distances by water or the glaciers of the Ice Age. Alluvial deposits on the inside of a river bend provide a simple example. In such cases there may be some organic matter present from the start, and soil solution may be easily available from the water-table, so that the process of soil building can go on rapidly. In alluvial deposits, whether they be marine, at lake bottoms or along rivers, there is usually a sorting out of particles brought about by the currents, so that if the deposits later become consolidated into sedimentary rocks the resulting material tends to be more uniform both chemically and in particle-size than that derived directly from igneous rocks.

After the preliminary mechanical breakdown of the rock, we may distinguish three processes in soil formation operating simultaneously, and it is the interaction of these which largely determines the nature of the soil. Their relationship is shown diagrammatically in Fig. 2.6. First there is the gradual **chemical weathering** of the mineral particles, in which the predominant feature is the removal of the stronger bases into solution. Secondly there is **leaching**, a process in which the soluble compounds resulting from chemical weathering are washed or leached out of the surface layers of the soil. The extent of the leaching is greatly influenced by local climate, and of course by the time for which the process has been going on (a reflection of the maturity of the soil). As the residual mineral complex tends to become increasingly acid with chemical weathering, excessive leaching not only de-pletes the soil of available bases but also affects its acidity. The third process is that of **humus formation** from the decay of plant and animal remains. This is greatly influenced by the acidity and drainage of the soil, which govern the type of vegetation and the population of soil arthropods and micro-organisms which it can support. Checks on normal bacterial decay due to unfavourable soil conditions commonly lead to a further increase in acidity from compounds present in the partly rotted humus. Leaching also plays an important part here in the removal of some of the soluble substances resulting from organic decay. The question may be asked: what becomes of the soluble material that is leached away? A proportion of it may be absorbed by plant roots growing deeper in the soil, to be built up into plant substances and later returned to the surface layers as constituents of the fallen leaves which

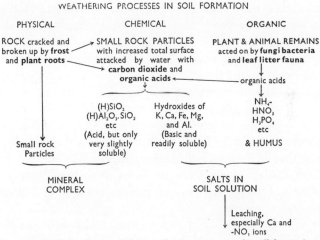

FIG. 2.6. Outline scheme of the weathering processes in soil formation

form the raw materials for humus. Under some conditions a large part of the leached material may be deposited in the lower layers of the soil, usually in the form of compounds not available to the roots of plants; this is especially true of iron. Doubtless the bulk of it is carried away with drainage water, eventually reaching rivers and the sea, but it is noteworthy that drainage water, especially that with a relatively high content of bases, may serve to enrich some of the soils through which it passes on its way. This process, which is the reverse of leaching, is known as **flushing**, and small areas of flushed soil with verdant grass are not uncommon around springs in upland regions. Enrichment on a much wider scale may take place where rivers draining from limestone hills pass into fenland before they reach the sea, as in

East Anglia. There is an interesting case of this occurring in prehistoric times, shown by the studies of pollen found in peat deposits on Sedgemoor (Godwin, 1951). The sudden appearance in Bronze Age peat of a preponderance of pollen from lime-tolerant species, where there had been typical heath vegetation, coupled with the discovery in the peat of 'corduroy' log-roads built by Bronze Age men across the swamps, provide evidence that the area was flushed with alkaline water from the Mendip Hills during a period of greatly increased wetness. The soil water was evidently so enriched as to change the nature of the vegetation.

It will be evident from this outline of the processes in soil form-ation that a natural soil, undisturbed by agriculture, will show a marked tendency to develop into layers or **horizons** of different composition and appearance. This **stratification** takes on charac-teristic patterns, and the study of soil profiles gives a clear under-standing of the development and nature of different soil types. The term **soil profile** is used to denote the whole range of strati-fication occurring in a soil, down to and including the subsoil. Horizon nomenclature has been changing, and the system followed here is that of Avery which has been adopted by the Soil Survey of Great Britain. Details can be found in the Soil Survey Field Handbook. Interpretation of soil profiles is based on recognition of five major processes:

1. *Weathering of the parent rock material* which underlies the soil proper, giving recognizable changes in colour and structure from the unaltered material. Unaltered subsoil material is designated as a C **horizon**, while altered material overlying it forms the (B) **horizon**. Where soils are derived from material which has been transported by glaciers or rivers, there will be nothing corres-ponding to parent rock beneath.

2. *Decomposition of leaf litter* and its admixture with the mineral matter in the upper layers of the soil. Unincorporated litter on the surface is designated as L (undecomposed), F (decomposing) or H (well decomposed humus) **horizon**; the upper layer of mineral soil, well mixed with organic humus is termed the A **horizon**.

3. *Leaching of colloidal and soluble material* from the region of the A horizon (and $L, F,$ or H layers) by percolating rainwater. Leached A horizons formerly referred to as A_2 are now called E (eluvial) **horizons**; they are recognised by their bleached, ashen colour.

4. *Deposition of colloidal or soluble material* (leached from the upper layers) in the B horizon. '(B)' denotes *only* weathering alteration; if there is also enrichment from deposition of leached material (evident from the darkening in colour), it becomes a B **horizon** (no brackets). Deposition sometimes takes place in sharply defined dark bands or pans of humus or iron, referred to as B^h or B_{Fe}.

5. In waterlogged soils beside rivers or with otherwise *impeded drainage* anaerobic conditions result in bluish grey reduction compounds of the minerals present in the (B) or C horizons, usually mottled with rusty patches. The mottling is especially marked along root channels, suggesting that local oxidation of Fe'' to Fe''' may be concerned. Such soils are known as **gley** soils: gleying may be evident in a (B) or C horizon, then referred to as $(B)_g$ or C_g.

The profile formed will be largely independent of the type of parent rock, but characteristic for any given set of climatic conditions, as the changes tend towards a state of approximate equilibrium in the mature soil. A mature soil should not be regarded as a static system, as the point of equilibrium probably changes with the gradual progress of further chemical weathering and leaching.

The climatic conditions most affecting the course of soil formation can be summarized as the ratio **rainfall/evaporation** which determines the balance between downward and upward movements of water in the soil. The fertile soils in the south and east of Britain result from a climate with a fairly low R/E ratio. Comparatively light rainfall and warm conditions give moist, well-aerated soils favouring the bacterial decay of leaf litter to form a mild humus known as **mull**. This is rapidly broken down to release, as mineral salts, the elements required for plant nutrition; there is thus a quick turnover of these raw materials, and the soils can support a rich vegetation. As the rainfall is light, leaching out of the bases is not excessive, a state which favours both the fertility of the soil and the formation of mild humus. Such soils belong to the general type known as **brown earths**. Beneath the surface layer of litter the A horizon is comparatively rich in bases, including calcium and iron oxides (hence the colour), and contains a fair proportion of mull. Earthworms abound, and it is largely through their activities that the aeration is good and the mull well incorporated with the mineral matter. The (B) horizon in these brown earths differs little in appearance from the A horizon just

B

described, and no sharp dividing line can be drawn between the two. The A horizon is usually mildly acid; the (B) horizon is slightly less acid and tends to contain more mineral salts.

The soil types commonly found in Britain contrasting to the brown earths are the **podsols**.* These develop in the bleaker upland regions, especially in the north and west, where the R/E ratio is highest and downward movement of water in the soil predominates. They are cold, poorly aerated or sometimes water-logged soils in which the bacteria responsible for mull formation cannot flourish. The processes of organic decay are greatly hindered, and yield a sour humus with a highly acid reaction, known as **mor**. Intensive leaching, due to the high rainfall, leaves the soil depleted of bases, a state which is often aggravated by the initial poverty in bases of many of the rocks. This has a twofold effect on soil processes, for the high acidity inhibits the beneficial mull-forming micro-organisms and at the same time no bases are left to neutralize the organic acids of the mor. In extreme cases, decay is so slow that the mor accumulates as peat. Earthworms are usually absent, and the advantages of their aerating and mixing activities are thus lost. This lack of mixing doubtless accentuates the stratification which intensive leaching brings about in podsols. The topmost layer of mor, L, is sharply marked off from the mineral soil of the eluviated E horizon beneath, which has been leached free of practically all the bases as well as any colloidal humus or clay particles. With the removal of the iron bases, which are readily soluble in organic acids from the mor, the impoverished soil of the E horizon is left grey in colour, though the upper layer is often stained brown by organic substances percolating through from the mor above. The leached material is deposited two or three feet below the surface in sharply localized zones known as **pans**. There are often two of these; an upper one B_h, dark brown in colour, in which colloidal humus material predominates, and below this a reddish brown pan, B_{Fe}, of precipitated iron com-pounds. In time they may become consolidated into hard, rock-like layers (**'hardpan'** or **'moorpan'**) impervious to water and plant roots, when the soil above becomes waterlogged and can only support bog vegetation. These changes are shown in Fig. 2.7.

Soil types intermediate between the brown earths and the pod-

* From a Russian word meaning 'ash' — the colour of the E horizon.

sols just described may be regarded as stages in a series of development in which the trend is for all soils to reach the podsol state
eventually. Progress is accelerated by upland conditions, and the
tendency towards the development of hardpans and peat bog
formation may represent yet a further stage in the series. The
drier climate of the lowlands allows a period in the summer when
upward movement of the water in the soil predominates, tending
to undo some of the work of leaching. Also, these soils, because of
their lower acidity, can support a vegetation which maintains a

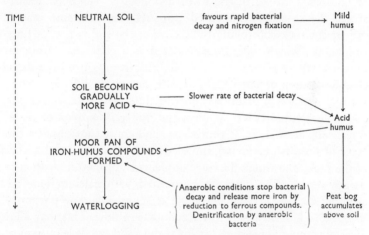

EVOLUTION OF ACID SOILS

FIG. 2.7. Interrelationships of changes in the evolution of acid soils.
(Adapted after Pearsall, 1950)

greater circulation of bases, especially calcium ions (leaf litter
from beech has been shown to contain more than five times as
much calcium as that from heather). Thus a state of approximate
equilibrium is reached in which the tendency towards podsol
development is so checked that progress becomes imperceptible.

It has been emphasized that climate (R/E) is the master factor in
the soil-building process, but we can see that the nature of the
parent mineral material must play a part in hastening or delaying
the podsolization of the resulting soil. Acid conditions must
obviously develop more readily in soils derived from siliceous rocks
already poor in bases, such as the Millstone Grits of the Pennines
or the Devonian Sandstones of Wales and the south-west. In the

south-east, where brown earths are the normal soil type, podsols with their characteristic heath vegetation have developed on coarse, porous sands favouring rapid leaching and already acid in tendency. In the same way a delaying effect is evident where soils develop from base-rich rocks, even in a typical 'podsol climate'. For example, the soils developed on basic volcanic ash rocks in Snowdonia show remarkably little tendency to podsolization, even under a rainfall of over seventy-five inches a year. But in interpreting soils of upland regions one must be alive to the possibility that profiles showing little zonation may in fact represent the *B* horizon of podsols from which the leached *E* horizon has been washed away; the so-called **truncated podsols.** The extreme case of the parent rock controlling the nature of the soil is that of limestone soils, which must be considered as a class apart from the brown earth — podsol series. Weathering results from the direct chemical action of water charged with carbon dioxide, and where the rock consists of practically pure calcium carbonate scarcely any insoluble residue is left to form the mineral complex of a soil. Such soils are known as **rendzinas**, and consist merely of a thin layer of humus overlaying almost directly the weathered surface of the rock, into which the roots of plants can penetrate with comparative ease. The whole is permeated with calcium ions and strongly alkaline. If the limestone contains a fair proportion of impurities, a mineral soil will accumulate in time. This may allow a certain amount of surface leaching, and a soil can develop eventually which is acid near the surface but strongly alkaline in the deeper layers. The vegetation found on soils of this kind is an odd mixture of shallow-rooting heath plants with deeper rooted lime-loving species.

The fertile layer which supports the growth of roots consists of a framework of mineral particles of varying sizes, ranging from colloidal dimensions to pebbles. The following American classification is widely adopted:

Gravel, particles more than 2 mm. in diameter (rock fragments)

Coarse sand, particles 2–0·2 mm. in diameter ⎫ (often
Fine sand, particles 0·2–0·05 mm. in diameter ⎬ largely
⎭ silica)

Silt, particles 0·05–0·002 mm. in diameter
Clay, particles less than 0·002 mm. in diameter
    (colloidal particles, chiefly alumino-silicates)

Any soil contains a mixture of particles of differing sizes, but where those of one particular group predominate the soil takes the name of that group. A good general mixture is called a **loam**, and highly cretaceous clays are known as **marls**. Humus particles are of course also present, mostly in a state of fine division.

The relative proportions of the different groups or **fractions** present in a soil is determined by *mechanical analysis* of the soil. For accurate determinations a somewhat elaborate process based on sieving is used, but useful information, especially when comparing two soil samples, may be obtained from a rough separation by sedimentation. All that is needed is a measuring cylinder (200 c.c. is a convenient size) and distilled water. After removal of any large stones, soil from the sample is put into the measuring cylinder up to the 30 c.c. or 40 c.c. mark, and made up to 200 c.c. with distilled water. This is shaken up until an even suspension is obtained and then allowed to settle. As the larger particles will obviously sediment out more quickly, a rough separation of the fractions is obtained, and their relative proportions can be estimated from the widths of the various bands which are clearly seen in the settled material. Colloidal matter remains in suspension and the larger humus particles float at the surface. It is, of course, difficult to distinguish accurately between fine sand and silt. By plotting a graph of the depth of settled material (in terms of divisions on the measuring cylinder) against time, further interesting information may be obtained.

One of the most important attributes of fertile soil is that it is *not* simply an assortment of closely packed particles, but has a definite crumb structure. This means that each soil particle is an aggregate of tiny mineral and humus fragments cemented together. As these aggregate particles are relatively much larger than the fragments which form them, and are themselves honeycombed with pores, both the size of the soil pores and the total pore space are greatly increased (see Fig. 2.8). This is borne out by the fact that clay soils commonly have a much greater *total* pore space than sandy soils, in which the crumb structure is poorly developed. One of the main aims in the cultivation of agricultural soils is to produce a good crumb structure or tilth, with the result that the pore space in such soils may reach sixty per cent of the total volume. We have already seen the vital importance of pore space

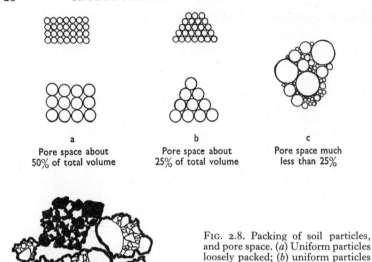

a	b	c
Pore space about 50% of total volume	Pore space about 25% of total volume	Pore space much less than 25%

d
Crumb structure: pore space may be as high as 60% of total volume

FIG. 2.8. Packing of soil particles, and pore space. (*a*) Uniform particles loosely packed; (*b*) uniform particles closely packed; (*c*) particles of various sizes closely packed; (*d*) crumb structure

to the well-being of plant roots, and it is not therefore surprising that a soil loses fertility if the crumb structure is destroyed.

Although the manner of crumb formation is not fully understood, various factors are known to contribute to it. For example, the secretions of soil bacteria and of plant roots themselves have been shown to play no small part in cementing mineral fragments together. The gels resulting from partial dehydration of soil colloids are also thought to have a cementing action; it is of interest here that a high proportion of colloidal humus fosters a particularly good, open type of crumb structure. Another kind of 'cementing', giving a very stable crumb structure, is achieved by electrostatic attraction between colloidal particles and ions bearing opposite charges; the flocculating effect of lime on clay soils being of this nature. E. W. Russell has explained the aggregation that occurs in dry soils in terms of negatively charged colloidal clay particles bonded to bivalent or tervalent cations by chains of water molecules orientated like bar magnets in an electrostatic field. Initially both the clay particles and the cations have their

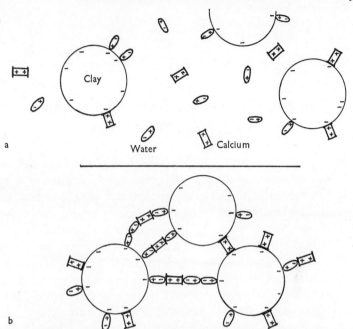

FIG. 2.9. Diagram illustrating the bonding of clay particles by electrostatic charges as the soil dries. (a) Wet soil; (b) dry soil

'shells' of orientated water molecules, and as the soil dries out the
shells are brought together to form an open framework holding
the clay particles together (Fig. 2.9).

Within the network of tiny, tortuous channels forming the
porous system of the soil is held the soil solution, with its teeming
population of bacteria, protozoa and moulds. It does not fill all
the available pore space, for a part is occupied by the soil atmosphere; the relative balance between these two will be subject to
wide fluctuations according to rainfall, drainage and the activities
of plant roots. The plant roots may occupy an appreciable volume
of the soil, and the root hairs growing between the soil crumbs
become fused inseparably to their surface, probably through the
formation of colloidal gels. Earthworms are also important inhabitants of this fertile layer of the soil; there may be more than
50,000 per acre. Their burrows help to ventilate the soil, and their
feeding habits result in a valuable mixing of the humus and
mineral fractions.

3

Variations in the Soil Environment and the Reactions of the Roots

Now that we have a general picture of the formation and structure of typical soils, we must consider how the factors of the soil environment can vary, and how root systems are enabled to meet the problems arising from these variations.

The basic pattern of growth of the root system, and of any other underground organs, is determined by heredity, and is characteristic of the plant species or variety. Thus one could say (though Floras are often strangely reticent on these points) that a particular species forms a tap-root, or a fibrous root system, or a rhizome bearing adventitious roots at the nodes. Closely related species, or even varieties may differ markedly in their manner of growth beneath the soil. Thus buck's-horn plantain (*Plantago coronopus*) forms a tap-root, but *P. lanceolata,* a common garden weed, has a fibrous root system; yellow flag (*Iris pseudacorus*) has a rhizome with adventitious roots growing along it, but the culti-vated Dutch iris a corm with a ring of adventitious roots at the base. But within this basic pattern, commonly adaptive in itself, there can be variations in behaviour which may hold much interest for the ecologist. The behaviour of root systems under different conditions offers a field of study which has been much neglected in Great Britain, perhaps for the simple reason that any detailed examination of root architecture is apt to be such a laborious business. But the very neglect of this field of investigation makes work in it all the more rewarding because there is so much to be discovered. A few practical suggestions for field work may not come amiss.

For any but the smallest plants the first step is obviously to dig a trench beside the roots under investigation. Here we may sometimes take advantage of sections of soil already exposed in

sand-pits, quarries, road or railway cuttings. It must be remem
bered, however, that unless these are *freshly* exposed, the behaviour
of the roots may be misleading, as the conditions of aeration,
drainage, etc. at the exposed surface will be abnormal. The exposed surface should therefore be cut back further, before the
examination of the roots is begun. The greatest difficulty lies in
removing the soil from the root branches. This can never be
achieved completely. The best technique is to wash away the soil
with a gentle stream of water from a hose, but this can seldom be
applied in the field. If there is a stream or pond really close at
hand, a garden syringe of the type that draws up its supply through
a rubber side-tube might be used. Failing washing, the best
method is to scrape the soil away, bit by bit, preferably when it is
dry and crumbly.

Root branches should be drawn to scale as they are exposed.
For this it is useful to have a folding square, made from four laths
bolted together at the ends; the square is divided into a grid of
suitable mesh by thin strings stretched across and threaded through
holes drilled in the laths at the appropriate places. This is placed
against the exposed soil face with the top lath horizontal, and the
string grid is used as a guide in drawing the root system on graph
paper. A thirty-inch square frame, with a string grid of three-inch
mesh, is a convenient general size, but for greater accuracy, or for
use with small root systems a one-inch mesh may be found desirable. Some saving in the problem of the strings getting into a
tangle when the frame is folded may be achieved by having a
permanent set of strings in one direction only, and completing the
grid square a row at a time by means of two temporary strings
held in position by bulldog clips.

The alternative method to excavating a trench is suitable only
for quite small root systems. It consists simply of digging up a
block of earth in which the entire root system is included, and
transporting this back to the laboratory, where the soil can be
washed away and the root branches floated out in water over a
glass plate.

By excavating root systems in this way, we can see not only the
general patterns of growth characteristic of each species (in just
the same way as one can recognize a tree by its branching pattern)
but also the differences in depth of growth, richness of branching,

etc. resulting from variations in the soil environment. It remains
to try to correlate these differences with the effects of particular
factors of the environment. Although variations in temperature
are less severe in the soil than for shoots growing above the ground,
they can still be considerable, and the roots must also meet variations in the composition of the soil atmosphere and the soil
solution. A close interaction exists between the effects of these
factors, making it difficult to consider any one of them in isolation.
This should be kept in mind when, for the sake of convenience of
arrangement, the variations in the soil environment are discussed
under the following headings:

(1) The soil atmosphere and drainage.
(2) Soil temperature.
(3) Soil water.
(4) The soil solution and its reaction.
(5) Soil organisms and humus.

It will be seen that the first two of these influence primarily the
growth and activities of the roots, while (3), (4) and (5) concern
more directly the supply of materials available for absorption, and
use by the plant as a whole.

(1) THE SOIL ATMOSPHERE AND DRAINAGE

As we have seen, the soil atmosphere occupies that part of the
space in the soil not taken up by soil solution, and the balance
between the two will fluctuate with rainfall and drainage. Prolonged heavy rain can lead to the displacement of much of the
soil atmosphere by water, but as this drains away fresh supplies
of air will be drawn into the soil behind it. Drainage and the soil
atmosphere are thus inseparably linked.

In a good soil some 20 per cent of the total volume (more than
one-third of the pore space) is filled with soil air, and, under almost
all conditions, this is kept saturated with water vapour by the
enormous total surface of the films of water covering the soil
particles. Desiccation, then, presents no problem except for roots
growing very near to the surface, but this does not mean that there
is always water available for absorption.

If the soil atmosphere is constantly moist, its composition in
other respects is very variable and differs greatly from that of the

air. To understand this we must remember that the soil is the scene of active respiration by teeming millions of micro-organisms, in addition to earthworms, soil arthropods and the underground parts of plants. It has been estimated that a soil of average fertility may produce as much as seven tons of carbon dioxide per acre, annually. The evolution of carbon dioxide from a soil sample can readily be demonstrated in the same way as is used to show the respiration of germinating seeds. In contrast to the aerial environment, however, the soil has no compensating gas exchange from photosynthesis, and ventilation is hampered by the small size of the pores. As a result, the soil is usually low in oxygen and contains at least seven or eight times as much carbon dioxide as the air; the precise composition depending upon the rate at which respiration is going on and the facilities for ventilation and gas exchange with the air.

The amount of organic matter present plays a major part in determining the rate of respiration, since upon this depend the soil micro-organisms and the earthworms, which take more oxygen and make a bigger contribution of carbon dioxide than the respiring roots. Soil temperature is also important, and its effects produce a seasonal fluctuation in the carbon dioxide content of the soil From a mimimum in winter this rises to a peak with the warmth of spring, after which it falls off a little during the summer (presumably owing to reduction in supplies of organic matter available for rotting down), and there is a second, smaller peak in autumn.

Soil ventilation must depend to a large extent on the *size of the soil pores* as distinct from the total pore space. Thus a sandy soil having a pore space occupying only about 20 per cent of its volume is much better ventilated than a clay-loam which may have more than 50 per cent pore space. In the clay-loam the pores are so small that diffusion is slow and the circulation may be blocked by water sealing in the soil atmosphere. Leaf litter, especially when the leaves lie flat, as those of beech, may further hinder ventilation. As much as 0·14 per cent of carbon dioxide has been recorded in the humus layer of beechwood soils; nearly five times as much as in the air. Rain-water, with its dissolved air draining through the soil may play part in replenishing the supplies of oxygen.

With increasing depth, the concentration of carbon dioxide may reach as high as 5 per cent, when it can exert an inhibiting effect on the growth of roots and the germination of seeds. Indeed, this narcotizing effect of high carbon dioxide concentrations may well be a major factor in preventing the germination of seeds buried deep in the soil. Brenchley (1918) has shown that, when arable land is turned over to grass, the seeds of many typical arable weeds will lie dormant in the soil to germinate after many years when opportunity is provided by the land being ploughed up again. The high concentration of carbon dioxide typical in grassland soils seems a very likely cause of this continued dormancy, though scarcity of oxygen for respiration may also be concerned.

In the case of actively growing roots it is even more difficult to sort out the effects of low oxygen content and excessive carbon dioxide in the soil. Both have a depressing influence on healthy respiration, which is essential not only for the growth of the roots, but also for efficient absorption of mineral salts. As a result, root systems in heavy soils with a high carbon dioxide content generally show stunted development in the deeper layers of the soil, but are richly branched near the surface. In stratified soils, root systems may show prolific branching in the better aerated layers; and sometimes vigorous growth can be seen along cracks or the burrows of earthworms where conditions of aeration are locally improved.

The concentration gradient of carbon dioxide with increasing depth may have a further effect in influencing the direction of growth of roots and rhizomes. It is thought that a chemotropic response of this kind may explain, at least in some cases, the way in which rhizomes continue to grow at an approximately constant depth in the soil, despite the surface of the ground being uneven. When we realize that the carbon dioxide concentration can rise to 150 times that of the air within the depth of a couple of feet, it is easier to understand that a mechanism such as this might control direction of growth. In some species the direction of growth of the rhizomes appears to be light-controlled, depending on the length of the vertical stem which is darkened by earthing up, (or experimentally, by covering it with light-proof wrapping). (See p. 132).

Having surveyed the general factors influencing the soil atmosphere and the effects that its variation may have on plant roots, we can now review the part it plays in particular types of soil.

At the one extreme are sand dune soils, in which the ventilation is so efficient that the soil atmosphere differs little in composition from the air. There are several reasons for this. In the first place, wind action results in a sifting of the sand grains, so that in any one locality they are remarkably uniform in size. As we have already seen (Fig. 2.8), the pore space between spherical particles of uniform size must, on theoretical grounds, be at least 25 per cent of the total volume; and will be much greater if they are loosely packed, as one would expect with the frequent movement in a sand dune. Salisbury (1952) in his admirable account of dune soils quotes figures of around 40–45 per cent for actual determinations of pore space in sand dunes in different localities. Furthermore, as the particles are coarse, the spaces left between them are large enough to allow easy ventilation and rapid drainage. Carbon dioxide production is restricted owing to the limited supplies of organic matter and conditions inhospitable for soil organisms. Rapid circulation is promoted by the rise and fall of the watertable with the tides, acting like a giant breathing mechanism, and by convection currents caused by the surface sand becoming heated on sunny days. It is clear from this that the soil atmosphere in dunes is favourable to rapid respiration and growth by plant roots, provided only that supplies of moisture are adequate. Salisbury has shown that the root systems of many quite small dune perennials may be very extensive, both in depth and lateral spread. Incidentally, dune soils provide an ideal medium for easy excavation and the study of root systems.

Another type of soil in which the atmosphere is particularly favourable for root growth is that of chalk downs. Here again there is good drainage, though the texture is nothing like so coarse as that of sand dunes. As we have seen, the soil proper (a rendzina) may reach no further than three or four inches below the surface, but plant roots are able to grow deep into the porous chalk itself (though their excavation is difficult). Here dissolved carbon dioxide produced by respiration is removed by combining with the calcium carbonate of the rock, to form soluble calcium bicarbonate, which percolates away through the rock with the drainage water. The fact that carbon dioxide can never reach excessive concentrations is doubtless an important factor contributing to the vigorous growth and branching typical of plant roots in chalk soils. In experiments carried out by Salisbury with roots of the same species grown in calcareous and non-calcareous soils, he found an average increase in the aggregate length of the root branches of $3\frac{1}{2}$ times in horseshoe vetch (*Hippocrepis comosa*) and $1\frac{1}{4}$ times in field fleabane (*Senecio integrifolius*).

Leaving the more extreme cases of good aeration we have the general range of ordinary soils in which the state of the soil atmosphere depends on a balance between the rate at which respiration is proceeding, and the facilities for ventilation allowed by soil texture and water content. There is generally an increasing concentration of carbon dioxide with depth, though this may be irregular in stratified soils and the gradient even reversed where there is a rich layer of leaf litter. Reference has already been made to the high carbon dioxide content of grassland soils with their densely packed surface layers and close tangle of shallow roots below. This condition may be accentuated by excessive treading down by stock, as around trees or through gateways, or by constant wear along footpaths. Bates (1935) has shown that apart from the mechanical injury to the plants, puddling, and consequent local waterlogging of the surface layers, is the primary cause selecting the characteristic flora of field footpaths, in which *Poa pratensis* is the most tolerant and regular member. In experiments which he carried out by making footpaths through bands of different grasses, he found that some species, such as *Agropyron stolonifera* and sheep's fescue (*Festuca ovina*) can stand treading quite well under dry conditions, but will not tolerate puddling.

At the other extreme are the completely waterlogged soils of marshland, bogs, swamps and aquatic habitats, where oxygen deficiency presents the main problem. Unless specially adapted to these conditions, plant roots are usually killed, even by quite short periods of exposure to them. Bracken is a particularly good example, and serves as a useful indicator of better drained areas when one is walking in boggy moorland districts. L. W. Poel (1951) in a study of soil aeration in relation to bracken, grew plants in culture in waterlogged mud and showed that without aeration wilting of the fronds occurred within 24-hrs. and the death of the plant followed. In continuously aerated mud the plants were able to survive and grow new fronds bearing sporangia. The roots of most plants are probably capable of some anaerobic respiration, but it seems doubtful whether the energy liberated in the process can be utilized to any extent in the syntheses necessary to maintain life and growth. Furthermore, anaerobic respiration results in the excretion of various toxic substances rendering the soil sour. Most of these are readily oxidizable, so that they will soon disappear under conditions of improved aeration. Plants like cotton-grass (*Eriophorum* spp.), rushes (*Juncus* spp.) and the various aquatics which can thrive in stagnant, waterlogged soils are characterized by well-developed air passages in the tissues through which oxygen from the air, or from photosynthesis in the leaves can diffuse to the underground parts.* The root systems are typically shallow, forming a dense mat two or three inches below the soil surface. Waterlogging, of course, leads to profound changes in the soil population of micro-organisms, and the effects of these changes on the acidity of the soil and its content of mineral salts will be discussed under Section 5 of this chapter.

The varying soil aeration requirements shown by different plant species must clearly be of great importance in competition under natural conditions. Only plants such as cotton grass, rushes, etc., with specialized physiology and structural adaptations for aeration of the roots, can tolerate the anaerobic conditions of moorland bogs, but they are enabled to grow there unhampered by competition from larger plants such as the heathers or bracken, which, on better drained soil would oust them. Under less-extreme conditions, tolerance of high carbon dioxide concentrations in the soil may determine whether a species can compete successfully, though it must be remembered that, with the gradient which exists in many soils, plants of varying tolerance may flourish side by side with their roots at different depths.

Some notes on the collection of samples of soil atmosphere are given in the Appendix.

(2) SOIL TEMPERATURE

Farmers often talk of clay as a cold soil, and indeed there is a lag in the spring growth of vegetation on clay soils sometimes sufficiently marked to be quite noticeable when passing to another soil formation on a train journey made in April. Phenological records of the time of first growth and appearance of early-

* Doubt has recently been thrown on the efficiency of 'aerenchyma' in facilitating the oxygen supply to roots growing in anaerobic conditions. It has been suggested that such roots overcome the problem by having fewer respiring cells per unit volume rather than better oxygen transport. (Williams and Barber, 1961).

flowering species on different soils in any locality can provide much interesting information on this point, and serve to emphasize the significance of soil temperature as an important ecological factor.

Generally speaking, the soil climate is equable so far as temperature is concerned. Day-and-night fluctuations practically disappear below about 2 ft., while seasonal variations become less evident, and show a marked time-lag in cooling after the summer and warming up again in spring. Extremes are unusual below the surface layers, and in Great Britain they are rarely large enough to cause direct damage to plants. Nevertheless, soil temperature plays an important part in the life of roots and underground storage organs. The primary function of absorption is directly affected, being so much reduced in cold weather that less water is *available* to plants in winter. This has doubtless been a vital factor contributing to the success of the deciduous habit in temperate climates. The pace of cell metabolism will obviously be influenced by soil temperature, and consequently the rate of growth of roots, and the mobilization of food reserves in seeds and underground storage organs. Both of these govern the availability of materials for growth of the shoots, and their effects will be particularly noticeable in the spring, though root growth must of course continue throughout the life of the plant and may be critical in time of drought, when the roots must push on and exploit new areas of soil if the water supply is to be maintained.

Delay in spring growth of underground buds, as those of rhizomes, corms or bulbs, through slow warming up of the soil will be accentuated if the perennating buds are at any considerable depth. For this reason plants with perennating buds at ground-level form a relatively higher proportion of the vegetation in cold northern climates. They are protected from desiccation and excessively cold winds by the insulating blanket of snow during the winter, but when this melts in spring growth can start at once, and none of the short northern summer is wasted. This relation between climate and the life-form of plants has been studied by the Danish botanist Raunkiaer, whose classification of plants according to their life-forms will be discussed in Chapter 5.

Other inhabitants of the soil, especially micro-organisms, are also influenced by soil temperatures, and the decay of humus is brought practically to a standstill during the winter. This is borne out by figures obtained for seasonal variations in the content of inorganic nitrogen in a medium comparable to the soil — the water of the Grand Western Canal (Fig. 3.1). Although the water is choked with dead vegetation in late autumn, there is scarcely any increase in the inorganic nitrogen from November to March, but as soon as the temperature of the water begins to rise in the spring, bacterial decay recommences, with a sharp increase in the content of inorganic nitrogen in the water, even though absorption

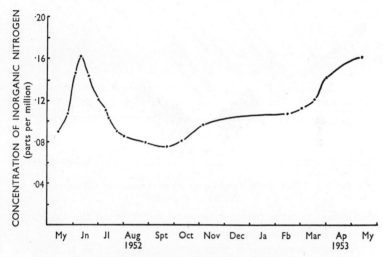

FIG. 3.1. Seasonal changes in the concentration of inorganic nitrogen in the water of an overgrown canal

by the new spring growth must have begun (*Blundell's Science Magazine*, No. 8).

In a rich soil, once the temperature creeps high enough for bacterial activity to get under way, then the heat produced will itself have some slight warming effect. This is unlikely to play an important part in natural soils, though a slightly earlier spring emergence has been claimed on very rich agricultural soil, such as ploughed-in leys, and the principle is, of course, used by gardeners in making hot-beds for raising early lettuces. Investigation of the state of affairs in a heap of grass cuttings showed the temperature to rise to $52°$ C. in a matter of four days.

Other purely physical effects of soil temperature are in the loss of water by evaporation from the heated surface layers, and its

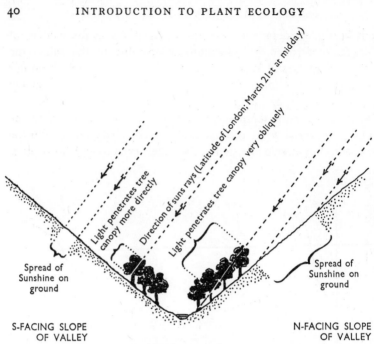

FIG. 3.2. Effect of slope and aspect on the amount of radiant energy
reaching the ground

deposition at night inside loose sand dunes when the circulating
air is cooled below dew-point.

We must now consider what factors come into play to make one
soil 'warmer' or 'colder' than another, under comparable weather conditions. These may conveniently be grouped under two headings:

(i) *Factors affecting the absorption of radiant energy from the sun.*
Slope of the ground and aspect (or direction in which the slope
faces) are here of prime importance. Obviously, on a north-facing
slope the radiant energy from the sun is spread over a larger
relative area, and hence 'diluted' as Fig. 3.2 shows, while a slope
with a southern aspect, with the sun's rays striking it at 90° at
midday, will receive the maximum radiant energy. However
obvious the principle, the magnitude of the difference is not so
easily realized. Figures for total solar insolation throughout the
year, given by Ashbel (1947) show an increase from 182 kgm.
calories per sq. cm. on level ground to 212 kgm. calories per sq.
cm. on a south-facing slope of 30° inclination, while for a north-
facing slope of the same inclination there is a drop to 102 kgm.

calories per sq. cm. South-facing slopes of 5° or 10° may have increased solar insolation equivalent to a difference of hundreds of miles in latitude. This effect becomes greater with increasing latitude, and in Great Britain is most marked in spring, when soil temperature has its greatest influence on growth. A careful comparison of vegetation and habitat factors on the north- and south-facing slopes of a steep valley, carried out in the spring of 1957, showed that this difference in soil temperature can have interesting secondary effects. Not only was the soil on the south-facing slope about 5° C. warmer throughout April and May (readings taken at a depth of about 3 in.),drying out more quickly and giving a higher rate of growth to the ground vegetation, but the beech trees on that slope came into leaf nearly a week earlier than on the north-facing slope. Evidently in the warmer soil their roots could absorb the necessary water more rapidly. Fig. 3.3 taken from a paper in *Blundell's Science Magazine*, No. 12, shows graphs of temperature recordings on the two slopes, and light readings at ground-level beneath the beech trees. The earlier breaking of the beech buds on the south-facing slope is reflected in the earlier fall in the light values. Differences in the soil moisture content on the two slopes, besides accentuating the temperature differences (through the effect on thermal capacity), seemed to influence the progress of humus decay, there being far more organic matter in the colder and wetter soil on the north-facing slope.

In the United States orchards are sometimes deliberately planted on north slopes, as the delay in the appearance of the blossom which results, may save some of the damage from late spring air-frosts. The chilling effect of cold north-east winds must also be taken into account in considering aspect, but its influence on soil temperature is far less than the check that it imposes on the growth of the shoots.

The amount of radiant energy actually absorbed by the soil must in turn be affected by the soil colour and vegetation cover. Mulching is an accepted gardening practice with the definite object of checking overheating of the surface layers of the ground, and leaf litter must have a similar effect in a woodland clearing where the sun's rays penetrate to ground-level. It is a simple matter to test such points experimentally, using shallow boxes with light sand or black soil (both should be equally dry), or a

SOIL TEMPERATURE

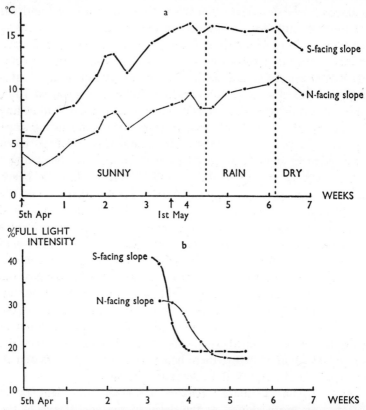

FIG. 3.3. (*a*) Comparison of soil temperatures (3 in. below surface) on north-facing and south-facing slopes of a steep valley in spring. (*b*) Relative light intensity beneath beech trees on the same slopes. Note the earlier breaking of the beech buds on the south-facing slope.

thin covering of mulch such as grass cuttings, and recording the temperatures at intervals throughout a sunny day by means of thermistors or thermometers with the bulbs buried to a constant depth (about 1 inch). It should be noted that soils with a good covering of vegetation or leaf litter, while absorbing less radiant heat, will also tend to *lose* less at night, when they will be warmer than bare soils.

(ii) *Factors affecting the actual warming up of the soil.* Water has a higher specific heat than any other substance. By contrast, the specific heat of the rock fragments forming the mineral particles of a soil is around 0·2. If we ignore other soil constituents, this means that a given amount of heat will raise the temperature of 1 cub. ft. of dry soil (with 50 per cent pore space) through three times the range that it would the same *volume* of soil when water-logged. We can now see why the water content of a soil has such an important bearing on soil temperatures, especially in spring, and why it is that clay is a 'cold soil'.

It is worth while describing a small scale experiment carried out in February and March 1952, with the aim of confirming these points (*Blundell's Science Magazine*, No. 8). Two deep wooden boxes, with holes for drainage, were filled, one with heavy clay and the other with light loam, and themselves sunk into the ground side-by-side. In each box eight sprouting corms of cuckoo-pint or lords-and-ladies (*Arum maculatum*) were planted, and a thermistor buried at a depth of 5 in., approximately the 'working depth' for these particular plants. The plants were watered in dry weather to eliminate any risk of growth being hampered through drought. They were allowed ten days to establish themselves, then, with the onset of warmer weather, soil temperatures in the two boxes were recorded daily at 8.30 a.m and 5.30 p.m. and the elongation of comparable leaves of plants in clay and loam was measured. The results are summarized in Fig. 3.4. It will be seen that after the night the loam soil was as cold as the clay, or even a trifle colder, but that it warmed up during the day to reach temperatures at least 4° or 5° C. higher than the clay. This was reflected in the growth of the leaves, some of those on the loam showing four or five times the rate of elonga-tion of those on clay. Admittedly, elongation of leaves is not by itself a very sound criterion of growth, but it does here reflect the rate of uptake of water by the roots, and also the rate of mobiliza-tion and translocation of food supplies, factors which govern the speed with which the plant can establish itself again as an efficient machine for photosynthesis.

Apart from the effects of its high specific heat, there is yet another way in which water content can affect the rate of warming up of the soil. What we have discussed so far concerns only the absorption of heat by the surface layers. As none of the constitu-ents of the soil can be regarded as good conductors of heat, and radiation below the surface must be negligible, effective trans-

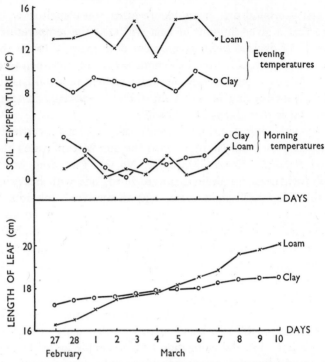

FIG. 3.4. Comparison of soil temperatures in clay and loam in early spring. The higher day temperatures in the loam are reflected by a higher growth rate of leaves of *Arum maculatum* in this soil, as shown in the lower graph

mission to the deeper layers can take place only by circulation of soil air, or the passage of warm drainage water. The establishment of natural convection currents will not be favoured, as the heating is from above, though in some exceptional cases, such as steep sand dunes, they may make an important contribution. Thus, transmission of heat to the lower layers is slow, even in a light soil, while in a heavy clay it must be further hindered by the high content of static water and the lack of free circulation of the soil air.

(3) SOIL WATER

Rainfall, of course, provides the immediate source of the water in the soil, though in some cases plants may derive their supplies from a high water-table, or even from dew. But the fate of rainfall on different soils and their capacity to store water in times of drought may vary considerably. Here we may learn something from research work into the fate of irrigation water, a precious commodity in dry climates.

Let us suppose that heavy rain comes to a very dry loam soil, where the vegetation cover has reduced the water content to about 9 per cent. If after two or three days we were to dig a trench and determine the water content of soil samples taken from various depths, we should find that the topmost layers would be evenly moistened (say 22 per cent water content) to a depth of about six or seven inches. Here, there would be a sharp dividing line, below which the soil remained just as dry as before. Had there been double the rainfall, the surface layers would have been no wetter, but the moist layer with about 22 per cent water would have reached to twice the previous depth, the soil below remaining dry as before. Let us now suppose that there is a prolonged drought. The vegetation growing in the soil continues to take up water, losing most of it in transpiration to the air, until the soil around the roots is down to a uniform 9 per cent. The plants then wilt permanently, and no further water is removed from the soil. This hypothetical case summarizes the findings of much experimental work, notably that of Conrad and Veihmeyer (1927 and 1929) in the United States. Two important conclusions can be drawn from it. The first is that, for practical purposes, *soil water may be considered to exist in three different categories:*

(i) *Hygroscopic water.* In the example just described, the last 9 per cent of the water in the soil was not available to the plants growing there, but was held by forces which the suction pressure of the root hairs could not overcome. It would have made little difference what species of plants were growing in the soil (whether they had a high or a low osmotic pressure); at just about the same moisture content permanent wilting would set in. This unavailable water — the percentage remaining when plants growing in the soil wilt permanently — is approximately a constant for any particular soil, and is known as the **wilting coefficient** of the soil (9 per cent in this example).

If a sample of soil is dried completely in an oven, and then left exposed to moist air, it will take up a certain percentage of moisture, although it will not *look* wet; this percentage is known as the **hygroscopic coefficient**. Experimental determinations show that these two constants are of the same order for any particular soil, though the wilting coefficient is always rather larger (some American workers claim a fixed relationship:

(hygroscopic coefficient = wilting coefficient × 0·68).

From a practical point of view it is the water unavailable to plants that matters, and this is commonly referred to as hygroscopic water, although it will be seen that this is really an approximation to the true meaning of the term.

(ii) *Capillary water*. After the rains had come to the dry soil in our example, the moisture content of the upper layers rose from 9 per cent to a uniform 22 per cent, and the ground would look wet. This extra 13 per cent was held by the soil in that it would not drain through, but was readily available to plants: it is called **capillary water**. The full 22 per cent, including the capillary and hygroscopic water is often called the moisture-holding capacity of the soil.

(iii) *Gravitational water*. With continued rain, the upper layers will not hold more than 22 per cent moisture against gravity, and any excess drains through to the lower layers as **gravitational water**. If the lower layers themselves become wet to their full moisture-holding capacity through any hindrance to drainage, then the soil would gradually become waterlogged, the water tending to occupy the whole of the pore space. Waterlogging at a considerable depth in the soil is common, and the level at which it occurs is called the **water-table**. This will tend to rise nearer to the surface in winter and subside again during the summer; close to rivers it will be permanently high.

Thus from our example we can say that the loam concerned had a moisture-holding capacity of 22 per cent, approximately 9 per cent being hygroscopic water and the remaining 13 per cent capillary water. These characteristics vary considerably in different soils; a light, sandy soil may have only 3 per cent hygroscopic water and about 13 per cent capillary water, while soils rich in colloidal material, either as clay or humus, may show values such as 13 per cent hygroscopic water and 27 per cent capillary water. (The percentages are normally expressed on the basis: weight of oven-dry soil = 100.)

The second conclusion to be drawn from the example described

is that *movement of water in the soil is restricted.* The existence of a
sharp line of demarcation between dry soil and that wet to its full
moisture-holding capacity shows that there is no rapid movement
of water to achieve equilibrium, as was formerly supposed.
Neither, in a dry soil, will water rise up from a deep water-table in
quantities large enough to support vegetation. This is of great
consequence to root growth, for it means that in time of drought,
once a root has 'exploited' the particular region of soil in which
it is growing, and reduced the moisture content to near wilting
point, then it must grow on into fresh areas if it is to obtain further
water supplies. The enormous development of roots in many
desert and dry prairie species is doubtless linked with this.

This restricted movement of water in the soil is better understood in terms of the modern concept of the soil pore spaces as a
collection of distinct 'cells', the water movement taking place
from one cell to another largely as a result of differences in vapour
pressure. Under very dry conditions the water is localized as tiny
rings just around the points of contact of neighbouring soil
particles, and any capillary movement is out of the question, but
if the soil is moistened these rings join up to give a continuous
water phase, rather like a network in that the air spaces too remain
continuous. With further increase in moisture content the air
spaces become discontinuous, and good aeration is hampered.

Once the water phase becomes continuous there is the possibility of some capillary movement, but this appears to be much
less than was formerly thought. It seems that quite considerable
forces are needed to move the water, and in investigating this
American workers have developed the concept of **capillary
potential** as an expression of the suction force or reduced
pressure required to draw water from any specimen of soil. It is
expressed as the height (in cm.) of the water column needed to
produce this pressure. A simple technique for the determination
of capillary potential is described in the Appendix. It will be seen
that, in the soil, water will tend to move from a region of low
capillary potential to one in which the value is higher; also that as
the moisture content of the soil diminishes its capillary potential
will rise. Studies of capillary potential with changing water content
in different soils have helped to clarify some of our ideas about
movements of soil water. In this work, because of the large range

of values, it has been more convenient to use the logarithm of the capillary potential, and this is known as the pF of the soil. (For example: if a pressure deficiency corresponding to a column of water 100 cm. high were needed to draw water from a particular soil sample, then the pF of that soil would be 2.) Many soils with 30 or 40 per cent water may have a pF of 2, and at wilting point the pF of any soil is about 4·2, though soils with high water-holding capacity will contain more water at this pF value.

The differences in water-holding capacity shown by different types of soil depend on particle size and especially the proportion of colloidal matter present, either as clay or humus. Clearly, the greater surface of the finely divided particles gives the greater capacity for water retention, but the proportion of hygroscopic water, unavailable to plants, will also be higher. Some of this is involved in gel formation with the colloids, and the rest is held by surface forces which the suction pressure of the root hairs cannot overcome. In a natural soil we should, of course, expect some degree of stratification, so that the water-holding capacity at different depths may show striking variation. This may modify the general pattern of root growth in individual species; it can also mean that different species growing side-by-side, but rooting at different depths, may have their roots in quite different 'soil climates' — a factor which should be taken into account when studying competition.

Modification of the pattern of root growth is achieved in two ways:

(i) *By hydrotropism*, in which the direction of growth of root tips and branches is turned towards regions of the soil with a higher moisture content. The response is greatest when the roots are in a dry soil to start with, while above an optimum soil moisture content there is no hydrotropic response at all. Hence the working depth of a root system tends to strike a compromise between conditions of adequate water supply and good aeration. This compromise will be achieved in different ways depending on the hereditary growth pattern of the root system, its plasticity, and the tolerance or 'preferences' of the individual species. The practice of grafting choice strains of roses on to wild briar stocks, or the high quinine-yielding *Cinchona ledgeriana* on to stocks of C. *succirubra* is in part a recognition of the plasticity and tolerance

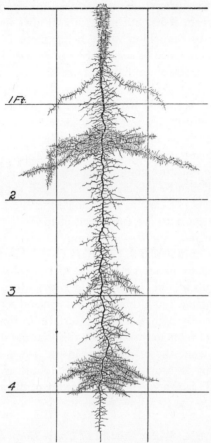

FIG. 3.5. Root of sugar beet grown in fine sandy loam soil, showing root stratification in the second and fourth foot where layers of clay were encountered

From Weaver and Clements: Plant Ecology, 1938, by permission of McGraw-Hill Book Co.

of the root system of the stock, which can adapt itself to grow vigorously in a variety of soils.

(ii) *By increase or checks in the vigour of growth and branching.* In a dry, well-drained soil, hydrotropism may prevent a root system from penetrating below a layer which, by reason of clay or humus, has a higher water content. Alternatively, the roots may penetrate deeper, but produce a rich network of branches in the moist layer (Fig. 3.5). While extreme dry conditions may check the progress of roots altogether, a relatively dry soil, with good aeration, will stimulate vigorous growth. This may take the form of growth in length of the main roots, but more often shows itself in a much greater degree of branching, there being more branches to the inch, and an abundance of secondary and tertiary branches. In a moist soil, by contrast, branching is reduced, and Weaver and Clements (1929) quote a case of maize plants grown in a moist soil having approximately half the root surface of similar plants grown in a relatively dry soil of the same type. Excessive water checks root growth, as may be seen in the shallow rooting common among marsh plants, but here the factor operating is poor aeration.

A favourable moisture content, especially if it is due to humus in the soil, will also be associated with a good supply of mineral salts, so that it is not easy to disentangle the effects of these two factors. Doubtless both are concerned in favouring the abundant root development in the surface humus of chalk soils, either in beechwoods or on downland. Salisbury (1952) has estimated that living roots occupy about 12 per cent of the volume of the first inch of a downland soil, as compared with only 1 per cent at a depth of 3–4 inches. But although in a downland soil the moisture content is generally highest near the surface, in times of drought the dense mat of roots may so deplete this that the shallow rooting species suffer, while those with roots in the deeper layers survive unharmed. This is brought out in the following figures sum-marized from V. L. Anderson's (1927) study of the water economy of chalk plants.

TABLE 1. *Water Content of Chalk Down Soils*

Depth (in.)	Water Content (per cent)			
	Mean	Minimum	Maximum	Range
0–3	37	8·6	61	52
6–9	27	11·5	36	24
12–15	26	10·7	34	23
27–30	27	15·3	35	19

In considering these figures it should be borne in mind that in the surface humus layer of the soil around 10–11 per cent water is held unavailable to plants as hygroscopic water, while deeper than three or four inches only about 4–5 per cent would be held as hygroscopic water. Thus, over the two-year period for which records were taken, there was always an adequate supply below the topmost 3–4 inches.

It is only in exceptional cases that dew assumes major importance as a source of water supply. Such are loose sand dunes, in which the air circulating through the action of convection currents may be cooled below dew point inside the dune, and pebble ridges such as Dungeness or Chesil Beach. In years of severe drought the sparse vegetation of pebble ridges may remain green, while that on neighbouring soils looks scorched and dead.

(4) THE SOIL SOLUTION AND ITS REACTION

In his book *The Living Garden* Salisbury (1945) has directed attention to the drain on soil resources which the constant removal of grass cuttings from a lawn involves. Adapting his figures to a modest lawn of 20 × 10 yards we see that the fresh weight of a single season's cuttings would amount to approximately 1,400 lb. This would be equivalent to about 200 lb. of dry matter, in which would be included some 6 lb. of nitrogen; 3 lb. of lime; and 2 lb. of phosphoric acid. Year by year this mineral drain continues on a close-cut lawn and commonly enough nothing is done to replace the losses, yet growth of the grass is maintained. But there comes a time when the soil is impoverished, and gradual changes set in, first in the species of grass, and then by the appearance of more and more weeds, like clover, daisies and plantains.

This illustration serves to emphasize two vital points: (i) the importance of the soil solution, from which growing plants must derive not only water, but their entire supplies of the elements nitrogen, sulphur, phosphorus, calcium, potassium, iron and magnesium, together with certain trace elements such as boron and manganese; (ii) that differences in the ability of soils to supply these elements in the soil solution lead (directly or indirectly) to changes in natural vegetation.

A field study which also brings out these points is the comparison of the colonization of an area of topsoil with a similar area of subsoil, such as may be left exposed when a new road-cutting is made. The difference in the rapidity with which the vegetation establishes itself as well as the differences in the colonizing species can be very revealing.

It is natural that we should want to know more about the soil solution, but in this we meet with immediate difficulty, for it is neither possible to extract the whole of the solution from a soil sample, nor even to obtain a representative specimen. We can estimate the overall concentration of dissolved salts from determination of depression of freezing point, but in order to find the proportions in which individual salts are present a sample of the soil solution must be extracted. This can be done by various techniques, such as suction, pressure, centrifuging or displace-

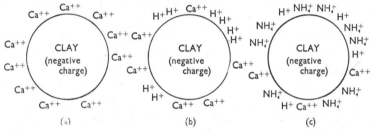

FIG. 3.6. Diagram to illustrate base exchange. (*a*) Soil saturated with calcium. (*b*) Soil not fully saturated with calcium. (*c*) The same soil as (*b*) after flushing with ammonium sulphate solution: most of the calcium and hydrogen have been replaced by ammonium ions

ment with liquid paraffin, but we can never tell how far the extract differs in constitution from the original soil solution, for it is well known that some of the constituents will be held back by colloidal adsorption or phenomena associated with base exchange. Base exchange depends on the fact that colloidal clay (and sometimes humus) particles in the soil behave like gigantic anions, the numerous negative charges at their surface holding positively charged cations (bases) (Fig. 3.6). These tethered cations can be replaced by others or exchanged; thus if a fertile soil is flushed with ammonium sulphate solution some of the NH_4^+ is retained, but a chemically equivalent quantity of Ca^{++} is displaced and will appear in the drainage water. The process could be continued until the clay anions were saturated with ammonium cations, and is also reversible. A soil can be saturated with any cation, but the acid radicles are not directly affected. An experiment sometimes used to demonstrate this property of soils is to flush two samples of well-packed clay-loam, one with a solution of methylene blue and the other with aqueous eosin. The eosin dye appears in the drainage water, but the methylene blue does not. The explanation lies in the chemical nature of the dyes; broadly speaking, methylene blue can be regarded as 'methylene-blue chloride', so that the part of the molecule responsible for the colour is held back by the clay anions, while eosin behaves as though it were 'sodium eosate', and only the base is held back.

Some cations are held more strongly than others, and, by indirect effects, buffering and allied phenomena, the same is true of acid radicles; thus in most soils the mineral complex tends to hold potassium and phosphate, while calcium and nitrate are easily leached out. We can only say that, although the soil solution must vary much in strength with the water content of the soil, it is very dilute indeed, probably of the order of 50–150 parts per million total dissolved matter. The phenomenon of base exchange is of more concern to agriculture than ecology, but mention should be made of the changes in mechanical properties in soils that it can bring about. In a fertile soil there is normally a fair proportion of exchangeable calcium held by the mineral complex, and this, as we have seen in Chapter 2, plays an important part in crumb formation. If such a soil is inundated by the sea, it may be changed overnight into a sticky, unworkable sodium soil in which crumb structure, with its attendant advantages of aeration and drainage have entirely disappeared. Reclamation of agricultural land thus flooded may take years (Chippindale, 1957), whilst the permanent existence of such conditions in salt marshes must set some problems for the growth of saltmarsh vegetation. Such plants must also be specialized in other ways to cope with the high osmotic pressure of the soil solution.

We saw in the previous chapter that the mineral salts in the soil have originated from gradual chemical breakdown of rock particles, and any soil receives constant, if meagre replenishment from this source. This is doubtless what keeps a lawn going for so long when we keep cutting it but never feed it. Once the mineral salts have become available to plants, they enter under natural conditions a constant circulation in which they are, in turn, built up into plant (and animal) substances and then returned to the soil again as decaying organic matter, to start the cycle once more. Thus the bulk of the mineral salts taken up by plants is derived from humus, and broadly speaking, all the ten major elements essential for plant growth, except carbon and oxygen, go through the same kind of cycle. But nitrogen occupies a unique position in the existence of an inexhaustible reservoir of the free gas in the atmosphere, while scarcely any is present combined in rocks. Micro-organisms which are able to synthesize their proteins using atmospheric nitrogen, bring more into general circulation in the same way as chemical weathering of rocks does for other elements.

While absence of humus, or conditions which prevent its complete breakdown in the soil will inevitably lead to shortage of all the essential elements in the soil solution, the effects of inadequate *nitrogen* are usually most noticeable, because this element is needed in larger amounts than others, and cannot be stored up in the soil effectively owing to the ease with which it is leached out. In this connection it is interesting to note that many heath and moorland plants, which grow under conditions of acute shortage of *available* nitrogen, show adaptations enabling them to augment the meagre supplies they receive from the soil. Gorse (*Ulex europaeus*) and bog myrtle (*Myrica gale*) have root nodules inhabited by nitrogen-fixing bacteria; sundews (*Drosera* spp.), butterworts (*Pinguicula* spp.) and bladderworts (*Utricularia* spp.) obtain nitrogen from the digestion of trapped insects. Indeed most of the insectivorous plants are either moorland inhabitants or epiphytes.

A rich supply of nutrients, especially nitrogen, promotes vigorous branching in roots, but no great extension, so that the root system tends to be more compact than it would be on a poor soil. Vigorous branching is also seen in the richer layers of a stratified soil (Fig. 3.5), the effect of water and soil nutrients here being hard to disentangle.

Of even greater interest to the ecologist is the element *calcium*, which affects the activities of roots both directly and indirectly, and has a profound influence on vegetation through the control which it exerts over the acidity of the soil.

The roots of many species, grown either in water culture or non-calcareous soils, show greatly increased development with the addition of relatively small amounts of calcium carbonate. While this increased development may be in part the direct result of available calcium ions, which are essential for plant growth, other secondary effects are almost certainly making a contribution. These are the effects of calcium on

(i) crumb structure and aeration (which we have already considered in Chapter 2);

(ii) uptake by the roots of other elements;

(iii) the reaction of the soil solution;

(iv) soil micro-organisms (which will be discussed under Section 5 of this chapter).

Effect of calcium on the uptake of other elements. It is well known that mineral salts are not absorbed by the roots *as such*, but as ions, and the protoplasmic membrane of the root hairs is selective in that it can let through one type of ion, but keep back another. This selective property is influenced by the composition of the soil solution, and here calcium, though by no means always beneficial, plays an important part. It appears that the presence of calcium ions in the soil solution antagonizes the uptake of phosphate ions, and of iron to such an extent that some plants (for example, broom) cannot make enough chlorophyll on calcareous soils. In the same way, the trace elements boron and manganese may occur in adequate quantities in the soil, but owing to the presence of calcium antagonizing their uptake, they may be unavailable to plants. The fact that only some species suffer from deficiency of these elements points to variations in the requirements, or perhaps in the selective properties in the protoplasmic membranes in the root hairs. Whatever the cause, it is clear that these species are not likely to compete successfully on calcareous soils.

By contrast, calcium in the soil favours availability of molybdenum, which is of particular importance to bacteria concerned with nitrogen-fixation, either in the soil, or in the root nodules of members of the Leguminosae. Many members of this family grow particularly well on chalk and limestone soils, possibly for this reason.

Calcium, and the reaction of the soil solution. By 'reaction' we mean the acidity or alkalinity of the soil solution, which is best expressed in terms of hydrogen ion concentration, for on this it really depends. Thus, a dilute solution of a strong acid, because

c

it dissociates more readily into ions, would produce the same hydrogen ion concentration as a much more concentrated solution of a weak acid. The concentration of hydrogen ions in a neutral solution is just under $1/10,000,000$, that is, $1 \div 10^7$ or simply 10^{-7} By convention, the exponent alone is used to express the hydrogen-ion concentration, ignoring the sign, and this is called the pH of the solution. At neutral point, then, pH $= 7$, and with increasing acidity the pH becomes smaller, while with increasing alkalinity (due to a preponderance of OH ions over the smaller concentration of H ions) the pH increases. It should be borne in mind that, as we are dealing with *powers* of 10, pH $= 5$ is ten times, and pH $= 4$ is one hundred times as acid as pH $= 6$. Approximate determination of soil pH is a simple matter, using a universal indicator.* It is customary to use a $1 : 2 \cdot 5$ soil-water suspension for this, which, because of the dilution, gives a result slightly more alkaline than that which must obtain in the actual soil solution.

The natural processes of respiration and decay in the soil tend to produce increasing acidity, due to the formation of carbonic acid and the various organic acids of humus and peat. The only base present in sufficient quantities to neutralize these is calcium carbonate, and herein lies its importance as a controlling factor, all the more so because both the bacterial flora of the soil and the roots of higher plants are so sensitive to changes in pH. In the case of roots the selective uptake of mineral salts is affected, so that the availability of important mineral nutrients changes with pH. Different species have pronounced 'preferences' for soils of different reactions, in which root growth will be most vigorous. This has been confirmed by their growth in a range of culture solutions differing only in their pH values; for one species maximum root development will occur at quite a different pH from that required by another species. Obviously, under natural conditions, each will compete best in soils having its optimum pH value, but will be ousted by other species where the pH is unsuitable. Such a mechanism underlies the distinction between many **calcicole** species, found only on chalk and limestone soils, and **calcifuges**, which cannot flourish on calcareous soils. As an example, the two bedstraws *Galium sterneri* and *G. saxatile* may be

---

* As supplied by Messrs. British Drug Houses Ltd. in their soil testing outfit.

quoted. Under garden conditions both can be grown on either a calcareous or a sandy soil, but in competition *G. pumilum* (a calcicole) will rapidly become dominant on calcareous soils, whilst on sandy soils it is ousted by the more vigorous growth of the calcifuge *G. hercynicum*. Experimental work suggests that only rarely is the distinction *directly* due to the plant's reaction to calcium in the soil. In other cases plants which we associate with chalk (for example, beech trees), will tolerate a wide range of *p*H, but demand good drainage.

The range of **tolerance** of some species is so narrow that their presence in natural vegetation gives a reliable indication of soil conditions. Examples of such *'indicator plants'* are sheep's sorrel (*Rumex acetosella*) growing in acid grassland, ling (*Calluna vulgaris*) on acid heath or moorland soils, and yew (*Taxus baccata*), when not planted, a sure indication to the geologist of a chalk or limestone outcrop. Isolated yew trees are often marked on 6 in. maps. Recently, a good deal of attention has been paid to mosses as indicators of soil and climate, particularly in forestry, and the Forestry Commission have brought out a booklet bringing this study within the reach of schools (Watson, 1947).

As calcium is readily leached out of the soil, stratification is usual, with the surface layers acid, and alkalinity increasing with depth. It is therefore not surprising to find shallow-rooted, acid-loving species growing side by side with deep rooting calcicoles. Salisbury (1952) has shown that some chalk species definitely benefit by having a part of their root system in the acid surface soil, and a part in the deeper alkaline subsoil. By comparison, the root development of plants grown in a uniform mixture of the two soil types was much retarded (Fig. 3.7). He suggests that the plants may derive their supplies of iron, manganese and boron from the surface layers, while the molybdenum is absorbed through the deeper roots.

The scale of losses through leaching of calcium is indicated by agricultural data, which estimate an annual drain of 2,000–3,000 lb. of lime per acre from calcium-rich soils. Salisbury (1921) has emphasized the importance of the changes in vegetation resulting from this leaching and the consequent increase in soil acidity. Natural regeneration of woodland has been inhibited, leading to extension of heath and moorland, especially at high altitudes,

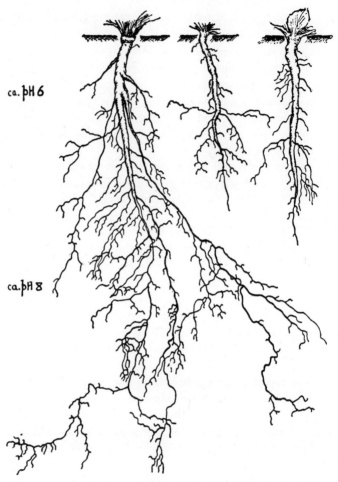

FIG. 3.7. Root systems of one-year-old plants of spotted cat's ear (*Hypochoeris maculata*). The two smaller on the right were grown in a uniform mixture (*p*H 7·8) of calcareous and non-calcareous soil. The larger plant on the left was grown in the same total volume of soil, but with the non-calcareous constituent (*p*H 6) above and the calcareous constituent (*p*H 8) below

From Salisbury Downs and Dunes, G. Bell and Sons, Ltd.

where leaching is most intense. The altitude limit of trees has thus been gradually depressed, though to a lesser degree in upland valleys on calcareous soils, where abundance of lime tends to hold up the process. His data obtained from the study of soils on a steep chalk down give a clear idea of what is going on:

TABLE 2. *Soil Conditions on a Steep Chalk Down*
(after Salisbury)

	Vegetation	*Soil Depth*	*Carbonates* %	*p*H
Summit of Down	Bracken	0–1 in.	0–0·02	5·1–5·4
		2–4	0·01–0·04	5·3–6·0
Halfway down slope	Chalk pasture	0–1	0·68–1·0	7·3
		2–4	1·1–3·2	7·3–7·4
Near base of slope	Chalk pasture	0–1	2·1–30	7·4–7·6
		2–4	25–65	7·6
Floor of dry valley	Chalk pasture with scrub	0–1	12·5	7·5
		2–4	20	7·5

We see in Fig. 3.8 that leaching is most intensive at the summit, where, if the ground is fairly level, a leached, acid soil will accumulate, supporting oakwood with bracken. On the shoulder of the down, erosion will prevent the accumulation of leached soil, and a typical beech hanger is common, with the roots of the trees embedded in the solid chalk. Below this, leaching decreases, and in the chalk pasture and scrub at the foot of the down eroded and leached soil tend to accumulate.

Among the various factors which help to delay the progressive

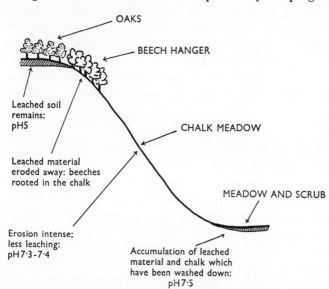

FIG. 3.8. Leaching and erosion on a steep chalk down. (After Salisbury)

acid trend of the soil as leaching proceeds one of the most important is the circulation of calcium through leaf litter. The calcium
in the leaves of trees is drawn by the roots from the deeper layers
of the soil, but redeposited at the surface when the leaves are shed.
The cutting off of this supply by the early felling of forests must
have played a part in the extension of heath and moorland. In a
beechwood, the contribution of calcium added to the surface of
the soil in leaf litter must be quite large, for the leaves have a high
calcium content; indeed, the surface litter may contain more than
the soil layers immediately below. But once the beech has to give
way to acid-tolerant vegetation such as pines, bracken and heather
the contribution is far smaller, as the leaves of these species are
relatively poor in calcium. Table 3 gives a comparison of the
calcium contents of some different leaf litters.

TABLE 3. *Calcium Content of Different Leaf Litters* (*Percentage
Dry Wt.*) (Figures quoted by Salisbury *loc. cit.*)

Beech 2·46	Bracken 0·83
Oak 1·70	Heather 0·44
Pine 0·99	

Earthworms perform a similar function in the addition of
calcium carbonate to the soil and humus passing through their
gut. This is deposited at the surface as the familiar worm-casts.
It must be remembered, however, that the worms must derive the
calcium from the soil in the first place, and only where it is being
taken from the deeper layers and brought up to the surface is any
useful contribution being made towards checking the effects of
leaching.

(5) SOIL ORGANISMS AND HUMUS

As we have seen from our illustration of a garden lawn (p. 51),
plants take a great deal of material out of the ground besides water.
To this there is the important addition of the carbon dioxide absorbed from the air in the photosynthetic process; for nearly half of
the dry weight of plant material is made up of the single element
carbon. Most important of all, photosynthesis results in the fixation of large quantities of energy, which can be set free again by
the oxidation of the organic compounds in which it is locked. The
energy fixed during the growing season by one acre of wheat has

been estimated as roughly equivalent to that stored in $3\frac{1}{2}$ cwt. of coal — an underestimate for the *total* energy fixed, since the figure is based on the yield and calorific value of the grain alone. Under natural conditions, then, leaf-fall and the dying back of the aerial shoots of herbaceous plants return much more to the soil than was ever taken from it. We have already seen that, starting from small beginnings, this process is an essential factor in the evolution of anything that can be called a soil.

This steady addition of plant remains, together with a modest contribution from animals, constitutes the raw material of humus, and, by reason of its abundant store of potential energy, supports a vast population of soil organisms living saprophytically upon it. Indeed, the investigation of leaf litter conveys so vivid an impression of the soil as a habitat crowded with living creatures that it makes a good practical introduction to soil work. Not only fungi, bacteria and protozoans, but armies of arthropods and nematodes are present. The food relationships of this teeming community are so complex that it is hard to disentangle all the steps in the breakdown processes. Fungi are clearly much involved in the early stages, but it seems that members of the litter fauna are also important as primary disintegrators (though they may also be fungus feeders). Their faecal pellets may form a considerable bulk in developing humus, and these in turn are utilized as food by such creatures as mites and springtails, as well as being attacked by fungi and bacteria. Many different species of bacteria are present, all in active competition for food substances, and these in turn are preyed upon by large numbers of protozoa. The numbers of organisms, even in one gram of soil, are so bewilderingly large that an indication of their total weight is perhaps more helpful. The fertile soil beneath one square yard may contain as much as 2 lb. of bacterial cells and correspondingly large weights of other soil micro-organisms, all concentrated in the humus layers comparatively near to the surface. Not only does this teeming population profoundly influence the soil conditions as an environment for roots, but also the balance between competing species of bacteria and fungi is itself liable to fluctuations sometimes with far-reaching results.

In the chain of reactions by which the breakdown of plant remains is achieved, particular species of fungi or bacteria are responsible for each stage. Thus, in the familiar steps of the nitrogen cycle

various fungi play a major part in the initial stages of conversion of proteins to ammonium compounds, and from here
the different nitrifying bacteria take over the further stages of
oxidation to nitrites and nitrates. Both the carbon dioxide from
respiration and the various organic acids formed in the early
stages tend to give the soil an acid reaction, which, if not checked
by neutralization with calcium, may hinder the work of the
nitrifying bacteria. If this occurs, further accumulation of organic
acids from the activities of fungi, which are less sensitive to pH,
may inhibit completely the action of bacteria responsible for
decay and lead to formation of a mineral-starved peat soil, typical
of moorland. Here we see again the importance of calcium in the
soil, for the maintenance of the correct balance of the bacterial
flora depends upon sufficient supplies to counteract the acids
formed in the processes of decay.

The efficacy of peat in arresting bacterial decay is shown by the
recovery of a Roman soldier, well preserved and complete with all
his equipment, from a peat bog on the Yorkshire moors. More
significant, if less spectacular, is the preservation of pollen grains
in peat over thousands of years. This has proved important to
ecology, for a study of the pollen grains preserved in successive
layers of peat formed year by year since the Ice Age has given us
much detail about the changes in vegetation and climate that have
taken place. As grasses and the forest trees forming the dominant
components of the vegetation all have wind-blown pollen, deductions from peat deposits apply to much wider areas than just the
immediate vicinity of the peat (see p. 195 and Fig. 10.5).

The pH of the soil and the supply of mild humus (**mull**) are the
most important factors controlling the earthworm population,
although, of course, earthworms cannot inhabit waterlogged soils.
Interesting confirmation of this was obtained by sampling fields
at different stages of crop rotation on a farm, and comparing the
earthworm populations with the known histories and state of the
soil in the fields (*Blundell's Science Magazine*, No. 6). There was
a close correlation between the populations of different fields and
their previous manurial treatment. Also, the planting of leys,
especially those containing leguminous plants, raised the earthworm population, and this was still further increased when the ley
was ploughed in. Incidentally, the estimates of earthworm popula-

tion per acre were far greater than the oft-quoted figure from Darwin; the highest was 395,000 per acre.

Good aeration and adequate water are further factors necessary for a healthy soil population. Excessively good aeration by over-cultivation of arable soils may lead to such vigorous bacterial action that supplies of organic matter are soon exhausted by complete rotting down. In contrast to this, the densely matted roots of grassland check aeration and so limit the pace of bacterial decay, explaining, in part, why grassland soils are so rich in humus. Anaerobic conditions of waterlogged soils or at the bottom of deep lakes may lead to peat formation in the same way as high acidity, by inhibiting the processes of decay.

In favourable conditions the nitrogen-fixing bacteria *Azoto-bacter*, which occur free-living in the soil, make a useful contribution to soil fertility through the extra atmospheric nitrogen which they bring into circulation by building it up into bacterial proto-plasm. This is, of course, an endothermic change, for which they derive the energy from the oxidation of carbohydrates in the humus. An addition to the soil of as much as 40 lb. of nitrogen per acre annually has been claimed for these organisms. Acidity or waterlogged conditions are alike in checking their activities. In the latter case, nitrogen-fixation by anaerobic species of *Clostridium* may take place to some extent, but as a rule it is more than counterbalanced by the activities of anaerobic denitrifying bacteria, which work in the opposite direction, breaking down nitrates in the soil to obtain their supplies of oxygen and liberating free nitrogen.

Water shortage will check the rate of activity rather than upset the balance of the soil population. In experiments carried out to investigate the course of production of inorganic nitrogen and temperature changes in a heap of rotting grass cuttings, it was noticed that both fell off rapidly below a certain water content, though there may have been other explanations (*Blundell's Science Magazine*, No. 10).

Temperature has an important influence in producing seasonal fluctuations in the rate of breakdown of organic matter, and hence in the supply of inorganic nitrogen available for plant growth. This is difficult to assess in the soil solution, but the investigation of the content of inorganic nitrogen in the water of the Grand Western

Canal throughout the year has already been referred to (p. 39).

As in the case of higher plants, trace elements may be vital for bacterial growth. It has been shown that one to three parts per million of molybdenum are essential for nitrogen fixation either by free-living *Azotobacter*, or the *Rhizobia* inhabiting the root nodules of leguminous plants. Another important aspect is that the availability of manganese for higher plants is to a large extent dependent upon the oxidation and reduction processes of manganese compounds carried out by soil bacteria.

Yet another factor lies in the type of raw material on which the soil organisms have to work. We have already seen how the calcium content of the leaf litter from different plants can vary (p. 60). Plants which can flourish on poor ground generally take less from the soil, and return less to it in their leaf litter. The resin contained in the leaf litter from pines and other conifers also delays rotting. As the soils on which they grow are usually too acid for earthworms, there is little mixing and the litter tends to accumulate as a dense surface mat. Decay proceeds very slowly indeed, taking some three or four years for completion, as compared with the usual two years for oak or beech leaves.

Enough has been said to indicate the vital part played by a healthy soil population in moulding the environment of plant roots, and the far-reaching results if the balance of this population is upset. Side-by-side with the free-living saprophytic fungi and bacteria there exist a number of species which enter into symbiotic relationship with plant roots. One can scarcely overestimate the contribution to soil fertility made by the nitrogen-fixing bacteria (*Rhizobia*) inhabiting the root nodules of leguminous plants. These have a free-living stage in which they are directly affected by soil conditions, and even when they are established in the root nodules, they cannot fix atmospheric nitrogen without traces of molybdenum absorbed from the soil. It should be remembered that isolated species in many other families of flowering plants besides the Leguminosae have root nodules inhabited by nitrogen-fixing bacteria; the riverside alders (*Alnus glutinosa*) provide a common example in which they are easily seen in roots from which the river has scoured the soil.

The other symbiotic relationship is that of the mycorrhizal fungi which grow in close association with the roots of higher plants. There are various kinds of mycorrhizas in which the physiological relationships between the symbionts differ. The most important ecologically are the ectotrophic mycorrhizas of many species of forest trees, as beech, birch or pine. Here the association is facultative, in that either partner can exist independently, and the fungus forms an external mantle covering many of the small branch roots and entirely suppressing the development of root hairs. Absorption, then, is carried out by the fungal hyphae, and a proportion of the accumulated ions is later released to the roots. It has been shown that pine seedlings make better growth when infected, and that under conditions of good oxygen supply, such as usually obtain in the mycorrhizal zone of the leaf litter and humus, the partnership of root and fungus is more effective than uninfected roots in absorbing ions of mineral salts. Many of these mycorrhizal fungi, it appears, cannot break down cellulose and lignin, but must absorb their carbohydrates in the form of sugars. In their free-living stage they must be at a grave disadvantage in competing with the unspecialized soil saprophytes. Although trees can grow without their fungal partners, there seems no doubt that they benefit by the association, and it could prove a critical factor in competition or regeneration of woodland. Many of the familiar toadstools of the woods are in fact the sporing bodies of mycorrhizal fungi, and it is often noticeable that particular kinds are always found at the foot of the same species of tree.

Another kind of mycorrhizal association (endotrophic) is found in all members of the heather family (*Ericaceae*) and all orchids (*Orchidaceae*). Here the plants cannot survive the seedling stage without the fungus, which grows in the intercellular spaces of the root cortex, penetrating into many of the cells. Both in the Ericaceae and Orchidaceae the physiological relationship between the host plant and its mycorrhiza is a complex one which has not been fully explained. In contrast to the Ericaceae orchid seeds are uninfected at the start and must 'pick up' their fungal partner from the soil if they are to survive. There may well be a connection between this precarious link in the life-cycle and the immense number of seeds produced by orchids.

Besides these saprophytic and symbiotic organisms there are also a number of soil-borne parasites, mostly fungi. Under natural conditions there is usually a nicely adjusted balance between parasite and host, so that the effects on vegetation are not conspicuous. Dramatic onslaughts are reserved for the golden opportunites presented to the parasite by man's interference in agriculture.

A side of soil microbiology about which we know very little as yet, but which is nevertheless of interest to the ecologist, concerns the part played by antibiotics in the soil. The fungi from which penicillin and patulin are obtained, and a number of other organisms producing antibiotics used in medicine are common soil saprophytes. One cannot escape the inference that these secretions may have a potent effect on competition and the general microbiological balance in the soil. In a few cases it has actually been shown that fungi, by the secretion of antibiotics, are able to check the vigour of fungal parasites attacking the roots or seedlings of crop plants. Failure of pines to grow on certain areas of Wareham Heath in Dorset has been traced to soil conditions antagonistic to the establishment of mycorrhiza: one suggested cause is the presence of antibiotics from various species of mould fungi (including *Penicillium* spp.) which abound in the soil. Antibiotic action may also explain some rather baffling cases of 'soil sickness', in which the breakdown in normal bacterial activity can be restored by partial sterilization of the soil, presumably removing the organisms producing the antibiotics. (See also p. 175).

4

Environment of the Shoot and its Functions:
(1) Light

There is overwhelming evidence from fossil records that plant life had its beginnings in water, and that the efficiency of present-day land plants is the outcome of gradual evolutionary improvements taking place over some 300 million years. The researches of the German botanist Hofmeister (1824–77) first demonstrated the life-cycle of the flowering plants (angiosperms) as the outcome of a series of progressive modifications of a simpler cycle, like that shown by the liverworts and ferns. This brilliant integration of the knowledge of the reproductive cycles in different groups of land plants gave tremendous impetus to the study of fossil forms (palæontology), and the evolutionary history of present-day plants from the early colonists of the land. Of course the Hofmeister series as studied in an elementary botany course (liverwort; moss; fern, *Selaginella*; Gymnosperm; Angiosperm) does not imply a direct line of descent, but it illustrates vividly the different stages in adaptation to terrestrial problems shown by different groups. As the efficiency of plants in competition must be largely dependent on how well they are adapted to their environment, a brief review of the main problems of land life is worth while at this stage.

One of the most urgent difficulties facing the earliest plant colonists on land must have been the scarcity of carbon dioxide in the atmosphere as compared with its relative abundance in solution in the water habitat. Without efficient assimilation of this gas no appreciable growth in size would be possible, and consequently adaptations giving an increased surface of wet cell wall for absorption of carbon dioxide from the air must have had considerable survival value. We see them in a primitive form in the photosynthetic hairs sunk in pits in the surface of liverworts such as *Conocephalum* and *Marchantia*, and they seem to have become stabilized as a basic pattern in the fronds of ferns, with their spongy mesophyll tissue so like that found in the leaves of flowering plants. The carbon dioxide content of the air does not vary appreciably with different microclimates, as do light and humidity, so we might expect that adaptations which meet this problem adequately will serve for all cases. Broadly speaking, this appears to be true, for one does not find much fundamental variation in leaf structure so far as internal surface for gas absorption is concerned. Indeed, even the photosynthetic tissue in the capsules of some mosses is built on much the same general plan.

Successful solution of the problem of carbon dioxide supply, however, inevitably increases the risk of desiccation, for the extensive surface of wet cell wall will lose water vapour rapidly to the dry air. This danger has been met by the evolution of a more or less impermeable cuticle, pierced by numerous stomatal pores, but there are many variations and special adaptations, corresponding to the wide range in atmospheric humidity in different habitats. These will be discussed in the next chapter.

Another consequence of life on land is the separation of the photosynthetic region of the plant from that concerned with the absorption of water and mineral salts, which must remain in darkness below the soil. With the prostrate growth of a liverwort thallus the distance of separation is so small that no special tissues for translocation are needed. Increasing competition on land would favour the evolution of stems, tending to carry the photosynthetic region above the level shaded by competitors, but this could not take place without a specialized vascular system to conduct water, mineral salts and organic compounds about the plant. Adequate provision for mechanical strength would also be essential. In avoiding shading from competitors, vascular plants with tall stems themselves become aggressive, by virtue of the shade they cast. Thus **light** has become a master factor in competition among plants on land.

Side-by-side with this progress towards more efficient vegetative form have gone the gradual changes in the reproductive cycle which are illustrated by the Hofmeister series. In the progressive reduction of the gametophyte generation, so ill-fitted to stand desiccation, and its retention in the megaspore, we see a trend leading to better and better adaptation to land conditions. With the evolution of the pollen-tube as a means of conveying the male gametes, complete emancipation from dependence on free water for fertilization is achieved. In this process the spore has lost its original function of dispersal of the species, since only the micro-spores are shed as pollen, but it is replaced by a 'new piece of biological machinery' — the seed. Being multicellular, this lends itself to adaptations for dispersal in many different ways, and it is at the same time highly resistant to extremes of drought and tem-perature. It is interesting to notice, in this connection, that while insects play so large a part in transporting pollen, there are very few adaptations in higher plants which secure insect help in seed dispersal.

From this brief review of the evolutionary background we can better understand why it is that angiosperms and gymnosperms form the bulk of the land vegetation. They are fully equipped for land life, and show evolutionary deployment in their adaptations to all kinds of differing habitat, while groups like the ferns or liverworts (the amphibians of the plant kingdom) are strictly limited to the habitats which they are able to colonize. Mosses, by reason of their *physiological* property of being able to stand desiccation and recover, have a wider range of habitat than a con-sideration of their structure would lead one to suppose.

We are now in a position to discuss in more detail the environ-ment in which the shoot grows. As in the case of the root, it is most conveniently regarded as the result of interaction of a number of different factors, which will be considered in turn.

Climatic factors   (1) Light
                   (2) Temperature ⎱
                                   (Chapter 5)
                   (3) Water ⎰
Biotic factors   (4) Effects of animals (including man)
                         (Chapter 6)
                   (5) Effects of competition from other plants
                         (Chapters 7–9)

LIGHT

Although light affects the growth and reproduction of plants in a number of ways, its most profound influence obviously lies in the supply of energy for **photosynthesis**. Day and night there is a continual expenditure of energy through the respiration of sugars, maintaining the normal vital functions of the plant and providing for additional growth. This must be met from the energy income from photosynthesis, or by drawing on the capital of stored food reserves. Since, at low light intensities, the rate of photosynthesis increases with brighter light; but respiration is not directly affected, the situation can be represented graphically as in Fig. 4.1. It will be seen that the **compensation point** is the light intensity at which the plant's energy income can just balance its expenditure in respiration. Below this light intensity it is living above its income, and can only survive by drawing on reserves of

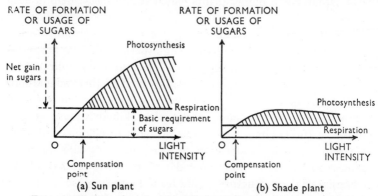

FIG. 4.1. The photosynthesis/respiration balance and light intensity. Note that a typical shade plant has not only a lower compensation point than a sun plant, but also it cannot make such good use of higher light intensities

stored food; above the compensation point there is a balance of unused sugar which is available for growth or can be put into storage. A very approximate estimate of the compensation point of a submerged water plant such as Canadian pondweed (*Elodea canadensis*) or one of the water crowfoots (*Ranunculus* spp.) is given by the light intensity below which bubbles of oxygen cease to evolve from the cut stem. The assumption is that in dimmer light photosynthesis still proceeds slowly, but all the oxygen evolved is immediately utilized in respiration. Another technique, applicable also to land plants, uses a number of stoppered test-tubes, each enclosing a leaf (or leaf-disc) over bicarbonate indicator (see Appendix), and exposed to a different light intensity. In the tube at compensation point for the plant under test, CO_2 usage and production are balanced, so the colour of the indicator remains unchanged.

Different species of plant vary widely in their compensation points. Both efficient performance at low light intensities (giving a steeper slope to the graph) and a slow rate of energy expenditure in respiration may contribute to give a low compensation point. The former may be the outcome of adaptations in leaf structure, such as the increased surface/volume ratio of bluebell leaves growing in deep shade, while the latter implies a thrifty plant making very slow growth. Conversely, plants having a high compensation point may show the vigorous growth that one associates with weeds. The difference may be obscured in many woodland herbs by the rapid spring growth from food reserves stored in the previous year; but few further leaves are added through the season. In contrast, the usual growth pattern of meadow plants is an increasing rate of leaf production as the total leaf area (earning capacity) builds up with the season — until production is diverted to flowering.

Generally speaking, plants having a low compensation point ('shade plants') cannot make efficient use of high light intensities, as some other factor, such as carbon dioxide intake, then limits the rate of photosynthesis. The reverse is usually true for 'sun plants', with their high initial compensation point. Thus, there is a working range through which any particular species can make the most efficient use of the available light, qualifying it as a sun or shade plant, and every possible intermediate stage occurs between extreme sun and shade species. Moreover, different individuals of the same species, and even different leaves on the same tree, show some variation in this respect, according to the local light-climate. The increased relative surface of the leaves of bluebells grown in deep shade has already been mentioned; another example is seen in the different structure of sun-leaves from the top and shade-leaves from beneath the crown of a tree such as a beech. The sun-leaves act as more efficient filters of the bright light, having two or three well-developed layers of palisade cells, while in the shade-leaves even the single layer may be poorly defined. It is easy to slip into a teleological explanation for this, and say that they have become adapted to exploit their own microclimates best, but more probably the *cause* lies in the different nutritional status of the leaves.

Although the distinction between sun- and shade-species is somewhat blurred by individual variations in differing micro-climates, the concept remains valuable in trying to interpret the distribution of plants growing in the field. It will be seen at once that the laboratory idea of compensation point as a definite light intensity can have little application under natural conditions, where the plant must make some excess of sugars during the day to maintain its respiration through the following night, and, as an added complication, the duration of daylight and darkness is constantly changing. A simple way of integrating all these variables is given by expressing compensation point as the percentage of full daylight which the species requires *in the field* to keep it alive. This is a reflection of the degree of shading that the plant will stand, and in areas which are more heavily shaded (that is, which receive a lower percentage of full daylight than the compensation point) the species will not be found at all. For this reason, the term **extinction point** is sometimes used, indicating the lowest percentage of full daylight in which a species is found under natural conditions. For its measurement all we need to do is take a series of light readings in the shadiest places where the species can be found, bearing in mind that specimens growing near their extinction point usually show signs of this by their puny size, absence of flowers, and sometimes long internodes and prostrate growth replacing a normally erect habit (for example, yellow archangel, *Galeobdolon luteum*). The measurements are taken with a light meter (see Appendix) at the average leaf-level of the species in question, and must be expressed as a percentage of the light intensity in the open *at the time*. This necessitates frequent readings taken in the open for comparison, and it is wise to choose a day of even, unbroken cloud, or else no cloud at all.

Most trees are naturally sun-plants, and under the shelter of their leaf canopies conditions of light, temperature and humidity are totally different from those in the foliage in the crowns. This different microclimate provides a biological niche in which smaller, shade-tolerant species can grow without competition from the more vigorous sun-plants of the open ground (Figs. 4.2a and 4.2b). Thus, a low compensation point might be regarded as a physiological adaptation fitting these plants for such a habitat, but the situation is complicated by the seasonal changes in shading which

FIG. 4.2a. Hazel Coppice—Spring.   Field layer of wood anemones
(*Anemone nemorosa*) in light phase

FIG. 4.2b.   Hazel Coppice—Summer.   Note absence of field layer

occur in deciduous woodlands, as we shall see presently. If the light penetrating the leaf canopy of the trees is sufficient to allow a tall growth of shrubs or herbaceous plants, there may be a third layer of extreme shade plants growing prostrate on the ground beneath these. In this way, the vegetation of deciduous woodland shows a characteristic stratified structure. In evergreen coniferous woods there is rarely enough light to support any ground flora at all. The same is true in tropical rain forests, where a vast number of different species of trees (mostly evergreen) spread their leaf canopies at different heights, and herbaceous plants can only survive as epiphytes, growing in pockets of soil on the branches of the trees with huge woody climbers, far up above the level of the ground.

The light-climate on the forest floor is obviously of great significance in connection with regeneration, for the tree seedlings must grow up in the shade cast by the parent trees. In Britain most of the natural types of woodland are more or less pure stands of a single species, or at the most, two or three species of tree, each casting a characteristic shade. There is an interesting general correlation between the degree of shade cast by different tree species and the shade tolerance of their seedlings, as shown by the following examples (though it must not be assumed that the relationship is always as simple as this).

TABLE 4. *Shading and shade tolerance in trees*

Species	Shade Cast	Shade Tolerance
Beech, Yew	Very dense	Very tolerant
Oak	Dense	Tolerant
Ash	Fairly dense	Fairly tolerant
Birch, Willow	Light	Intolerant
Scots pine	Dense	Intolerant

Thus regeneration of either beech or yew is possible in the dense shade of a beechwood, and indeed these two species are often found growing together, as also are oak and ash. Ash trees, however, cannot survive in competition with beech where the soil is sufficiently well drained for the latter to flourish. Scots pine affords an apparent contradiction, but it must be remembered that its winged seeds give very good dispersal, so that most of the seedlings will germinate clear of the shade cast by the parent tree. Natural pine forests, where they occur, are usually open in

FIG. 4.3. Beechwood—Summer. Note absence of field layer

FIG. 4.4. Lone pine showing self-pruning

character, thus allowing regeneration which could scarcely succeed where the trees are closely planted. Evidence of the intolerance of pines to shading is seen in the phenomenon of 'self-pruning' when they are planted in close stands. The lower branches are so densely shaded by the leaf canopy above, that they die at an early stage, giving a straight main trunk with a minimum of knots in the timber. Other factors doubtless contribute to the absence of side branches on old lone pine trees (Fig. 4.4). It is here suggested that the replacement of most of the needle spurs by male cones on older branches may be an important factor, since these branches must in time become so impoverished by reduction of leaf surface that they can scarcely maintain themselves.

Not only does the degree of shade vary for different species of deciduous tree, but also its seasonal *duration* is characteristic for the species. For the woodland ground flora this has even greater significance than the actual summer shading, for most of the species depend upon the spring light phase, before the tree canopy has developed, to manufacture the bulk of their year's food supplies (Fig. 4.2). The poverty of the ground flora in a beechwood probably owes as much to the early breaking of the beech buds as to the dense shade cast by the leaves throughout the summer (Fig. 4.3). By contrast, oaks, even when they are casting nearly as dense a summer shade as beeches, still allow a fairly rich ground flora because the leaves unfold some two weeks later in the spring (Fig. 4.5). This seasonal fluctuation in light climate leads, then, to selection of early-leafing herbs in the ground flora, which can make the best use of the light phase. The leaves of many of them have died back by the end of June, and the seasonal changes in the woodland floor are so striking that the terms winter aspect, pre-vernal, vernal, summer and autumnal aspect are often used to describe them (Fig. 4.6). Thus in its vernal aspect, the ground in an oakwood may be carpeted with bluebells, red campion, and woodspurge, whilst in the same area in summer one can see little beside a thin growth of bracken. With trees casting less shade, for example, birch or ash (Fig. 4.7), the seasonal changes in the ground vegetation are far less marked, as active photosynthesis is possible beneath the leaf canopy throughout the whole summer.

In woodlands where the dominant trees are late in coming into leaf (for example, oak, ash) and hence allow an appreciable light phase in the spring, species in the ground flora are not necessarily shade plants, but owe their success more to the seasonal peculiarities of their growth. This was well brought out in the intensive studies of the bluebell (*Endymion nonscriptus*) by G. E. Blackman and A. J. Rutter (1946–50 and 1954). In a series of carefully planned field investigations combined with physiological studies, they showed that this species has its maximum rate of photosynthesis in full daylight, that is, it could be a sun plant. With

FIG. 4.5. Oakwood — the result of clearing showing well-developed undergrowth with bracken and willowherb

falling light intensity the *rate* of photosynthesis declines, but the plant develops an increased leaf area relative to weight, giving it slightly greater efficiency. Down to about 70 per cent full daylight this compensates for the falling rate of photosynthesis; but at lower light intensities the growth rate and total weight of the plants are affected, resulting in decreased number and size of the leaves, and fewer flowers on the scapes. In deep shade many of the bulbs do not flower at all. Where the mean light intensity between early April and mid-June falls below about 10 per cent full daylight bluebells cannot survive, though there will be a time-lag in their disappearance if the shade has increased rapidly, as large bulbs can

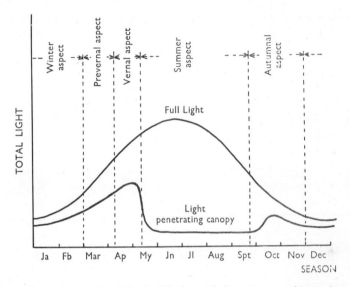

FIG. 4.6. Seasonal variation in light beneath the tree canopy in an oakwood. (Devon)

FIG. 4.7 Ash Copse—Summer. Showing well-developed undergrowth

survive for more than one season at light intensities below their compensation point.

Using a grid transect (see p. 114), Blackman and Rutter recorded relative light intensities and the density of bluebell plants (numbers per sample quadrat) in different types of woodland. By calculating the mean densities of plants for different light ranges (for example, 10–20 per cent; 20–30 per cent full light) they were able to show a rapid falling off in density below 40 per cent full daylight. Moreover, using statistical techniques beyond the scope of this book, they were able to establish that in shady woods, by far the greater part of the variation in bluebell density was to be accounted for by differences in light intensity. With increasing light, the influence of variations upon density was found to be less marked.

It is, then, scarcely surprising to find patches of bluebells flourishing on open ground; indeed, one would expect them to be commoner in such habitats than they actually are, since they can make good use of full light. But although bluebells do sometimes grow well in the open 'in their own right' (e.g. at Start Pt.), investigation often shows that such patches survive from the ground flora of felled woodlands, and may be gradually dying out. It appears that a number of factors other than light become important in open habitats; the plants are particularly susceptible to damage from grazing or trampling, and in competition with vigorous grasses they may not be able to obtain enough water. For this last reason, bluebells are found on open ground, such as the grass verges beside the road, more often in the west.

There is much scope for detailed studies of the nature of shade tolerance in woodland plants, and worth-while contributions can well be made by schools. Season and habit of growth, form of leaf mosaic, and adaptations of structure (such as those of bluebell leaves already mentioned, or the curious in-tucking of the palisade cells of elder leaves) may all play a part, as well as physiological properties, such as a low compensation point or ability to make full use of high light intensities.

In woodlands clothing steep hillsides, one often finds strikingly different ground vegetation on slopes facing in different directions. Different soil conditions, perhaps due to the angle of the bedding plane of the rocks, may contribute to this variation; but there can be no doubt that differences in the light-climate according to the direction in which the slope is facing, are most important. In seeking to analyse the habitat factors on two contrasting slopes in the Exe Valley in 1955, continuous readings of light intensity over the 24 hr. were taken on a model reproducing faithfully the angles of the slopes and the direction in which they faced (*Blundell's Science Magazine*, No. 10). A photo-electric cell connected up with a self-recording microammeter was used for the work, and the traces obtained are shown in Fig. 4.8. Slope *A* faced 8° north of west and showed an average inclination of just under 21°, while slope *B* faced 23° east of south at an angle of 27°; the records were taken during the first week in July. It can be seen that while slope *B* received 8 hr. of maximum light intensity, slope *A* had scarcely more than 2 hr., and, taking the total light received during the day

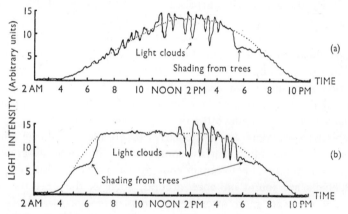

FIG. 4.8. Variations in light throughout the day reaching slopes of different aspect. Actual recordings made in first week of July: (*a*) Facing 8° north of west; slope 21°; (*b*) facing 23° east of south; slope 27°. Slope (*a*) received only about 78% of the light reaching (*b*). Notice how clouds not actually obscuring the sun act as reflectors and *increase* the light intensity

as proportional to the area beneath the graph, *A* received in all only 0·78 times that reaching *B*. For open grassland this would probably be immaterial, as one would expect carbon dioxide supply to limit the rate of photosynthesis in bright light, but beneath a dense tree canopy, through which only some 10–20 per cent of the light can penetrate, this difference may well prove critical for plants surviving through the summer in light intensities little above their compensation points. In the area studied, most of the west-facing slope was practically bare of vegetation, and in the absence of any striking differences in soil conditions, it was concluded that differences in light were responsible.

The incidence of sunflecks under different tree canopies and their effect on the light-climate is another aspect deserving of investigation. A fairly simple apparatus for measuring this is described by G. C. Evans (1956).

A special case in which local variations in the light-climate play a particularly interesting part is that of water plants. In the aquatic (or marine) medium, photosynthesis is scarcely restricted by shortage of carbon dioxide, as this gas is normally far more plentiful dissolved in water than it is in the air, so one would expect any variations in light to exert a greater influence on water plants. Shading of submerged plants arises not only from over-hanging trees and the floating leaves of water plants, but also from the absorption of light by the water itself. This will depend largely upon the amount of suspended matter, and its nature. Opaque mineral and humus particles are not likely to alter the *spectral quality* of the light, but abundant plant plankton near the surface will act as a colour filter, letting through predominantly green light, which is normally of no value for photosynthesis. During the summer of 1955 some preliminary observations were made, using a simple photo-voltaic light meter (see Appendix) and gelatine colour filters. In the first place it was established that the percentage total light absorption by clear water was unaffected by the intensity of the light. Tests for colour absorption by the clear water of a swimming bath indicated that red was absorbed most and blue least, but these results may have been influenced by the colour of the walls of the bath. In a disused canal the intensity of the blue light was found to fall off most rapidly with increasing depth, red light rather less rapidly, while the absorption of yellow and green was least of all. This is in agreement with what one would expect from the absorption spectrum of the chlorophyll in the plant plankton.

It is clear that this light absorption by water and suspended matter must play an important part in controlling the distribution of rooted aquatic vegetation. Beyond a certain depth (depending on the light requirements of the species) submerged leaves would be in light intensities below their compensation point, so that in colonizing fresh areas there must be enough food reserve in seeds (or rhizomes) to carry the newly formed leaves up to this critical depth. Obviously in deeper waters there will come a stage when this is not possible and all food reserves are exhausted while the leaves are still below the critical depth, so that colonization cannot succeed. Besides this effect, the seasonal shading cast by floating leaves of water lilies, pondweeds (*Potamogeton* spp.) and duck-weeds (*Lemna* spp.) will raise problems similar to those in wood-lands, which we have already discussed, but there is no marked 'spring flora' of submerged species, probably because of the slow warming up of the water at any depth (Fig. 10.2).

In the sea, light absorption by water must play an equally important part in controlling the distribution of deeper growing marine algae. The spores or zygotes have usually but little food reserve to carry the fronds aloft, so the sporelings must have an adequate light intensity from the start if they are to survive. Attention has often been directed to the general correlation between pigmentation and the depths at which different marine algae are found. The red algae commonly grow at greater depths than other groups, and it has been pointed out that the red pigment with the chlorophyll, being complementary in colour to the green-ish light penetrating to those depths, should enable the plants to absorb the prevailing light more efficiently. What matters is whether the plants can make better *use* of this quality light in their photosynthesis reaction, and there is good evidence that this is so, indeed, red algae may carry out more photosynthesis in blue-green light than they do in red light.

So far we have been concerned only with the effects of light on photosynthesis, and the consequences arising from them. Clearly the most important impact of the light factor on ecology lies here, but we must not overlook the *other ways in which light influences plant growth and reproduction*. The actual growth form can be considerably affected by both the quality and quantity of light. Thus, elongation of internodes may be particularly restricted by a high proportion of ultra-violet light, giving the compact habit of alpine rosette forms, while good leaf mosaics are largely the result of phototropic response to local differences in light in-tensities. Both these aspects are more appropriately discussed in

Chapter 7 as they have such profound effects on the competitive powers of the plants concerned. Light may also influence germination; the seeds of some species requiring light, while others are light-shy. This could lead to a correlation between good germination and particular types of soil surface, but the position is complicated by the effects of temperature in modifying the response of germination to light, and it is unlikely to prove important as an ecological factor. The responses in development of different species to long or short days (strictly speaking, short or long nights) will control the season of flowering and seed production, and may prove a barrier to interbreeding of closely related forms, as also may different times of opening the flowers during the day. The general light intensity of the habitat will influence the abundance of pollinating insects, an indirect effect which may well be associated with the early *flowering* as well as leaf formation in woodland species, that is, they are normally short day plants. Although the distribution of individual species will to some extent be influenced by their responses in the various instances mentioned, in none of them is light going to have a major effect on the pattern of vegetation. Those concerned with reproduction and dispersal will be further discussed in Chapters 8 and 9.

5

Environment of the Shoot and its Functions:
(2) Temperature and Water

In studying the effects of light in the last chapter, we have seen that the light-climate is like a fine mosaic of different micro-climates, subject to wide variations within the compass of a few feet. By contrast, temperature and precipitation are far less liable to local differences, but are the factors largely concerned with the geographical distribution of plant species. In discussing them we shall begin by reviewing the direct effects of temperature on the growth of the shoot. Much of what is said here applies also to the underground portions of the plant, but, as we have seen, soil temperatures are not subject to nearly such wide fluctuations as those experienced in the air, so that temperature effects loom less important beneath the soil surface. Some of the most important temperature effects are indirect, and closely bound up with the water balance of the plant; these will be considered second. Finally we must discuss the responses of plants to extreme short-age and to excess of water in the environment of the shoot.

DIRECT EFFECTS OF TEMPERATURE

It is a matter of simple physiology that the rates of respiration and growth in plant tissues increase with temperature to optimum values, after which they decline as the temperature becomes excessive (Fig. 5.1a). For any one species, both the respiration and the growth (cell division) curves follow the same general pattern (the temperature graph characteristic of enzyme action), but normally the optimum for respiration is higher than that for growth. Now, variations in the shapes of these curves and in the optimum temperatures peculiar to individual species, may be regarded as physiological adaptations enabling the plants to ex-ploit particular biological niches in their habitats. A flattish curve will mean a wider range of temperature tolerance than one which

climbs steeply to the optimum and falls away rapidly at temperatures above it (Fig. 5.1*b*). We have already seen how many woodland plants succeed by virtue of their early leaf formation, which
enables them to make good use of the light reaching the ground
before the leaf canopy of the trees unfolds. This means growth and
development of the leaves in cold weather, before the majority of
herbaceous plants have shown any signs of moving, that is, woodland plants of the pre-vernal and vernal phases require not only a
good reserve of stored food, but also low temperature-optima

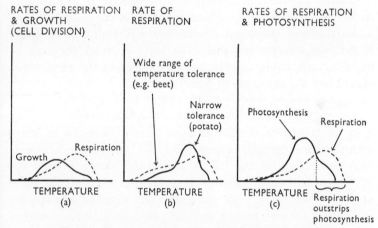

FIG. 5.1. Temperature/respiration relationships (see text)

for respiration and growth. Furthermore, the leaves must be able
to photosynthesize actively at relatively low temperatures, so the
temperature-photosynthesis curve, which follows the same general
pattern as the others, has also a low optimum. That for wood
anemones (*Anemone nemorosa*) Fig. 4*a*, has been estimated at
about 59° F., while some alpine species have been found to have
photosynthesis optima between 40° and 50° F.

The optimum photosynthesis temperature is normally much
lower than the optimum for respiration. This may have important
repercussions on the plant's photosynthesis/respiration balance,
for above a certain temperature the rate of income from photosynthesis will be falling off, while the expenditure in respiration is
still rising (Fig. 5.1*c*). The situation will be aggravated by warm
nights, when the consumption of food materials in respiration is

still further increased, and it will be seen that a temperature analogous to a compensation point may be reached, above which the plant cannot balance its expenditure in respiration. The principle is well illustrated in potatoes, which, although of tropical origin, are adapted to the cool climate of some 11,000–14,000 ft. in the Andes. Attempts to grow them in hot climates meet with failure because, even if the plants survive, there is no balance of food to form tubers. Such physiological adaptations must obviously be a check on the gradual migration of species to hotter climates.

Another aspect of metabolism directly affected by temperature is the rate of hydrolysis of food reserves, as can readily be demonstrated with starch and diastase in the laboratory. The nature of the food reserves and the temperature at which they can be mobilized effectively must be factors influencing the germination of seed and the sprouting of buds in spring.

TEMPERATURE-WATER RELATIONSHIPS

In the climate of the British Isles there is little risk of direct injury to plants through excessive heat, but high summer temperatures and exposure may check ill-adapted plants through the effect on their water balance. Transpiration can outstrip water uptake by the roots, even when adequate supplies of water are available in the soil, leading to temporary wilting, with closing of the stomata and a consequent halt in photosynthesis. The rate of transpiration will be governed by (a) external factors — the temperature and the relative humidity of the air, which are conveniently integrated as the 'evaporating power' of the air, and (b) internal factors — the most important being the state of the stomata and the permeability (if any) of the cuticle.

The evaporating power of the air shows wide variation with exposure or shelter, and at different times throughout the growing season (see Table 6, p. 95). It can doubtless be an additional potent factor restricting woodland species to the humid atmosphere which the protection of the tree canopy affords. Comparisons between the values in different parts of a habitat can easily be made using an atmometer, which measures the rate at which water is evaporated from an exposed surface of porous pot. Rapid determinations can be made on a basis of volume (distance travelled by an air bubble along a capillary tube in a given time).

These are apt to show considerable fluctuations with momentary breezes, and several readings should be taken in each locality. Loss in weight gives a more reliable indication over longer periods, if one can leave the instrument without fear of anyone tampering with it. Notes on various designs of atmometer are given in the Appendix.

The rates of transpiration of different species in a habitat are often compared using slips of dried cobalt chloride paper clipped on to the leaves between microscope slides (see Appendix). The papers (with fixed colour standards) are dried in the laboratory and transported in stoppered tubes containing calcium chloride beneath some wire gauze. The time taken for the paper to change from the standard blue to the standard pink when clipped on to a leaf surface is inversely proportional to the rate of transpiration under the conditions prevailing. Of course, both upper and lower surfaces of the leaf must be tested, and an expression of the total transpiration rate for both surfaces is given by the sum of the reciprocals of the times. The result is in arbitrary units, but allows a comparison between different leaves. The method has the advantage that it can be applied to leaves on the parent plant, though they must not be wet, and some types of leaf such as pine needles, those of heathers, rolled or very hairy leaves are unsuitable. It is however, open to criticism in that the rate of transpiration actually measured must be different from that of the leaves when freely exposed to the air, for the variable factor of the evaporating power of the air is replaced by one that is approximately constant — the absorbing power of cobalt chloride paper. Thus, in any studies where the reactions of different species to variations in atmospheric humidity are of importance, results from this method would be suspect, and it might be wiser to employ potometer methods comparing the rates of uptake of water by cut shoots (see Appendix).

In open habitats, the layer of air very close to the ground is liable to the widest daily fluctuations in temperature. Low-growing seedlings are thus particularly exposed to risks of damage from excessive heat or cold. Low temperatures pose some of the most critical problems for plant survival, and have resulted in the most profound responses to meet them. With fall in temperature all the metabolic processes; photosynthesis, respiration and growth are

D

brought practically to a standstill, and the tissues are liable to damage from cold. Direct injury from frost may result from ice crystals forming in the cell vacuoles and rupturing the protoplasmic envelope which encloses them. Precipitation of the protoplasmic proteins may also occur in the concentrated salt solutions remaining in the cell after effective withdrawal of water by freezing. Either of these would cause rapid death of the cells. As in the case of the high temperatures, however, the more general effect is one of desiccation brought about indirectly. On first sight, this does not seem to fit in with one's ideas of the temperate winter, but is explained by the fact that, with the falling rate of respiration water uptake by the roots is severely checked so that even a low rate of transpiration could scarcely be compensated. Under such conditions the effects of dry east winds would be devastating.

It is reasonable to suppose that higher plants were first evolved as evergreens in moist tropical climates with little seasonal variation in temperature or humidity. Adaptations which allowed colonization of unexploited temperate regions would have considerable survival value, as plants possessing them could grow with but little competition. Some groups, such as the conifers, have become adapted physiologically and structurally while still retaining the evergreen habit. The most widespread response, however, is the evolution of the deciduous habit, which has served also to meet the problems of a dry season in hot climates. It may be noted that even in a rainy tropical climate, with little seasonal variation, some of the trees shed most or all of their leaves at intervals leaving resting buds similar to the familiar winter buds we see here. It may well be that in the evolution of the deciduous habit a natural rhythm of this kind has become geared to the seasonal changes of colder climates. On the other hand tropical trees showing marked leaf-fall could be descended from ancestors adapted to seasonal changes, which have migrated back into the tropical rain forests. An important consequence of leaf-fall synchronized with seasonal changes is the biological niche exploited by the spring flora of woodlands.

If it is to serve for winter protection the deciduous habit involves: (1) the shedding of leaves before they become a liability to the plant, either through frost damage or by involving transpiration losses which cannot be made good by the roots; (2) the development of resistant buds (with varying degrees of protection against desiccation and cold) from which the following season's leaves can grow; and (3) the storage of food for these buds to draw upon until the plant can again be self-supporting from its new growth of leaves.

The onset of winter conditions may be very sudden, particularly in continental climates, so that it is essential that a successfully adapted species should have completed its annual reproductive cycle and 'prepared the way' for leaf-fall before the end of the growing season. The actual leaf-fall mechanism is not so much triggered off by any single environmental factor but appears to be initiated by a rhythm in auxin production which may be regarded as part of the plant's total reaction to its climatic environment. Most species are susceptible to damage by frost once the growing season has commenced, even though they may have considerable resistance in the dormant phase. The response of bud development to rising temperatures and lengthening days in spring, and the length of the summer growing period are thus factors limiting the geographical distribution of species according to the length of the frost-free period. This can vary widely with local topography and the proximity of the sea, even in the same latitude. The difficulties of introducing crop plants into new areas often bring home forcibly how nicely the plants are adjusted to their native climates. As an example may be cited the tung oil tree (*Aleurites fordii*) a native of the hilly country of the upper Yangtze valley in western China. The nuts of this tree yield a valuable varnish oil, for which it was introduced into the United States. But very careful choice of site was needed for in some areas of Florida and Louisiana where it was first planted there were severe losses from damage to the flowers by late spring frosts, yet when the trees were grown further south the *absence* of winter frosts upset their whole seasonal rhythm, making flowering and fruiting irregular and poor. Here we have a common phenomenon akin to vernalization, where a period of low temperature in the dormant stage of the seed or bud is beneficial in initiating the breaking of dormancy and hastening subsequent development.

The dormant buds which remain after leaf-fall must themselves be protected against desiccation and cold if they are not to be a liability to the plant. It is commonly supposed that the bud-scales (sometimes supplemented by sticky, resinous substances as in horse-chestnut or rhododendron) afford the protection against desiccation. This can easily be verified by comparing the loss in weight of twigs with normal and descaled buds. It also receives indirect support from the fact that the submerged turion buds of hydrophytes have no protecting scales at all. However, it is worth noting that the buds of some species may survive the winter after having their scales removed; also in other species, which may be quite hardy, this protection is anything but complete, for example, the ill-fitting scales of alder buds (*Alnus glutinosa*) or the naked buds of the wayfaring tree (*Viburnum lantana*).

Probably the most important protection is given by a natural reduction of the water content in the tissues of the bud. This is also the means by which resistance to cold is achieved, for it is idle to suppose that the insulating effect of a few bud-scales can keep the inner tissues warm when there is virtually no source of heat to conserve. Actual recordings of temperatures inside winter buds, made by means of a tiny thermistor inserted into them, show little difference from the air outside. On a sunny day the inside temperature may be about 1° C. higher, but at night it tends to fall below air temperature. Cold resistance must depend therefore upon depression of freezing point inside the cells. It has been shown that in some dormant buds most of the water is held imbibed by carbohydrate substances like pentosans and mucilages.

Beside this natural resistance, protection is afforded to many winter buds by the habit of growth of the plant concerned. While in trees they are held high above the ground, exposed to all the drying winds that blow, the winter buds of herbaceous plants may lie buried in the snow or beneath the soil where they will never experience the worst extremes of the weather. The advantage of such protection must have been a potent driving force in the evolution of herbaceous forms, for there exists a striking correlation between the severity of the winter climate (or dry season) and the preponderance of particular **life forms** in the natural vegetation. This is of fundamental importance as the characteristic features of the main vegetation types (such as deciduous woodland, heath, prairie, etc.) depend upon the life form of the dominant species.

The Danish botanist Raunkiaer (1934) devised a system of classifying the wide variety of life forms which has proved a valuable tool in ecological studies. It has the merit of wide application and is based on a simple principle — the position relative to ground-level of the perennating (resting) buds which carry the plant through the unfavourable season of the year, be it a hot, dry season or a cold winter. Trees, with their resting buds carried high above the ground, he regarded as the most primitive life form, being best adapted to the even climate in which it is thought the

FIG. 5.2. Examples of some of Raunkiaer's life forms. Perennating parts are shown in black. (a) phanerophyte (currant); (b) and (c) chamaephytes (sea thrift and house-leek); (d) and (e) hemicryptophytes (primrose and lady's smock); (f) and (g) geophytes (wood anemone and blue-bell). Diagrams to varying scales

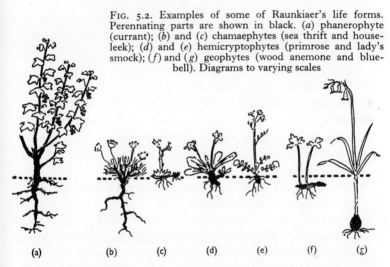

(a)   (b)   (c)   (d)   (e)   (f)   (g)

first flowering plants arose. Adaptation to winter climates or dry seasons of varying severity is achieved by life forms in which the perennating buds are borne closer to the ground or actually buried in the soil. The extreme case is represented by annuals, which survive only in the highly resistant form of dormant seeds. Raunkiaer's classification may be outlined as follows (Fig. 5.2):

PHANEROPHYTES — perennating buds borne well above ground-level, that is, trees and shrubs. This class is subdivided according to whether the plants are evergreen or deciduous, and the buds naked or protected with bud-scales. There is also a subdivision based on arbitrary limits of size, as buds borne nearer to the ground are less exposed to desiccation than those on tall trees:

MEGAPHANEROPHYTES — over 100 ft. high.
MESOPHANEROPHYTES — between 25 and 100 ft. } (symbol MM)

MICROPHANEROPHYTES (M) — between 6 and 25 ft., for example, cherry laurel, holly.

NANOPHANEROPHYTES (N) — between about 1 and 6 ft., for example, heather, gorse.

CHAMAEPHYTES (Ch) — perennating buds borne just above ground level; up to about 1 ft. in dense growth — under-shrubs like thyme, or herbaceous creeping or cushion plants, for example, sea thrift, *Stellaria holostea*.

HEMICRYPTOPHYTES (H) — perennating buds half-hidden in the surface of the soil e.g. dog's mercury, *Epilobium montanum*. This class also includes the rosette plants like daisies, plantains and dandelions.

CRYPTOPHYTES — with perennating buds buried in the soil or beneath standing water. The group is subdivided into:

GEOPHYTES (G) — perennating buds buried beneath the soil; plants with bulbs, corms, rhizomes, etc.

HELOPHYTES — marsh plants with perennating buds in waterlogged mud.
HYDROPHYTES — water plants with their perennating buds beneath the water. } (HH)

THEROPHYTES (Th) — Summer annuals, surviving the unfavour-able season in the form of seeds, for example, *Galium aparine*, *Veronica hederifolia*. Included also in this class are the winter annuals which germinate in the late summer and autumn, survive the winter usually as a rosette to flower, fruit and die in the following spring, for example, *Anthriscus neglecta*.

STEM SUCCULENTS (S) and EPIPHYTES (E) form further classes, as they are very characteristic elements of certain types of flora, even though they do not fall in line with the series outlined above.

Raunkiaer directed attention to the correlation between the type of climate and the 'biological spectrum' shown by the vegeta-tion, that is, the numbers of species in each of his classes of life form, expressed as percentages of the total number of species in the flora of the region studied. It is not these percentages them-selves that are significant, but rather their departure from those of the 'normal spectrum'. This 'normal spectrum' would, ideally, be that of the entire world flora, taken as a whole, but Raunkiaer had to use a working approximation to this, based on analysis of a random sample of four hundred species. Tables, taken from his work, shows the type of variation found in different climates.

TABLE 5

	Latitude	No. of species	The percentage distribution of the species among the life forms									
			S	E	MM	M	N	Ch	H	G	HH	Th
Normal spectrum		400	1	3	6	17	20	9	27	3	1	13
Franz Joseph Land	80° N	25	.	.	.	.	.	32	60	8	.	.
Spitsbergen	77° N	110	.	.	.	.	1	22	60	13	2	2
Denmark	56° N	1084	.	1	1	3	3	3	50	11	11	18
Samos (Greece)	38° N	400	.	.	1	4	4	13	32	11	2	33
El Golea (Sahara)	30° N	169	.	.	.	.	9	13	15	5	2	56
Seychelles Is. (Indian Ocean)	4° S	258	1	3	10	23	24	6	12	3	2	16

A high percentage of phanerophytes is typical of moist tropical climates. In the cool temperate climate of Denmark there are far fewer phanerophyte species (although they are still the dominant species of most plant communities) and hemicryptophytes become the ascendant class in the biological spectrum. In cold winters their resting buds are protected by the blanket of snow; but they receive the full benefit of the spring warmth as soon as the snow has melted. One can see that geophytes, with their buried buds, are at a disadvantage in colder climates, for the soil warms up so slowly in the spring that they are liable to waste a valuable part of the short growing season. They are in fact better represented in Mediterranean countries, where the hot, dry summer is the unfavourable season. In desert climates (El Golea) nanophanerophytes, taking the form of the characteristic grey, spiny bushes, are more numerous, while during the brief rainy seasons therophytes spring up and colour the whole landscape. Raunkiaer showed a similar series of changes in the biological spectrum of the flora with increasing altitude on mountains, and his system is also of value in analysing plant communities on a smaller scale.

Looking back to the third requirement involved in the deciduous habit as a means of perennation — food storage for the resting buds to draw upon when they develop — we see that this too is closely linked up with the life forms of the plants. Particularly in the case of geophytes and hemicryptophytes there are many special adaptations for storage in rhizomes, root and stem tubers, corms and bulbs. As we shall see in Chapter 7, these have also come to serve the purpose of vegetative reproduction.

Having reviewed the effects of temperature on the environment of the shoot, it is worth devoting a little space to suggestions about the kind of problems that can usefully be investigated.

In winter the governing factor lies in the limits of tolerance to cold and desiccation shown by the perennating buds. Observations on the soil temperatures at different depths (using buried thermistors), and on the temperature and evaporating power of the air at different heights above the ground will provide useful data about the environment. The insulating effect of snow cover is also of interest. Kimble and Bush (1943), quote an extreme case recorded in the United States, when the temperature of the snow surface was —27° F., while the temperature seven inches below the snow surface was +24° F. — a difference of 51° F. There is much scope in studies of the resting buds of different species; their internal temperatures; the changes in their water content and specific gravity from autumn to spring; their rate of water loss in dry air, with and without the bud-scales, and the effect on their survival of removing the bud-scales.

Spring provides a wealth of problems to be studied in the correlation between local differences of the environment and bud development. 'Frost holes', occurring when denser, cold air accumulates in hollows where it cannot drain away, may result in sharp localization of frost damage. On agricultural land the edge of the area affected may occasionally be seen as sharply as though it were drawn as a contour line. The local effects of a wall in frost protection are also worth careful study, and are sometimes shown clearly by the limits beyond which the buds of fruit trees have been damaged. Apart from these more spectacular, but less common, examples of frost effects, the delay in growth and flowering of species growing in 'frost holes' is of interest. Phenological records of the dates of first appearance of a number of species in different habitats, kept over a number of years, can yield useful data, especially if they can be correlated with weather records. The effects of north and south aspects of a slope on the rate of spring growth and the nature of the flora are well studied in a bank and hedge running approximately east-west.

Summer yields less local variation in temperature, but differences in the evaporating power of the air, as measured by atmometers, become more important. Figures quoted from Yapp (1910) by Adamson (1921) show how greatly this factor changes with the onset of autumn. The comparison is between a hot, dry June and a dry September with mist and heavy dew at nights.

TABLE 6

	Average daily loss in wt. of atmometer, at ground-level	
	June	*September*
In dense rosebay willow-herb, in a clearing	23·1	4·1
Bare soil under yews	37·2	7·2
Wood sanicle, in shade	46·8	9·5

EFFECTS OF WATER SHORTAGE OR EXCESS: XEROPHYTES AND HYDROPHYTES

Environments in which the water factor is at one or other of the extremes of the scale produce such striking responses in the plants which have become adapted to them that the names 'xerophyte' and 'hydrophyte' have been coined to describe such plants.

A xerophyte may be defined as a plant which is adapted to grow in dry places. As this is a problem which British plants seldom have to face, it may be more profitable to consider first the extreme case of desert vegetation. The popular idea of vast stretches of sand dunes without a sprig of vegetation in sight is by no means the rule among the world's deserts; more often there are brief, irregular rainy periods followed by months of drought, and under these conditions a fair variety of plants can survive. Root competition is so intense that bushes are widely spaced, sometimes almost with the regularity of planted orchards. Among this vegetation we find three distinct types of plant, each adapted in its own way to meet the problem of drought: the ephemerals, the succulents, and what may be called the drought endurers.

Ephemerals — These are the therophytes of Raunkiaer's classi-

D2

fication. They are small herbaceous plants which quickly spring up from seed with the onset of the first rain. They complete their life cycle in an astonishingly short time, flowering and setting seed before the ground has completely dried out again. Often they are so numerous as to transform the appearance of the whole landscape; hence such references in the Bible (Isaiah 35, 1) as 'the desert shall blossom as the rose'. The plants in their vegetative phase are in no way fitted to combat drought; they are prodigal with the water supply, on which they are entirely dependent. They pass the time of drought in the form of *seeds*, which are ideally equipped to withstand it. They are thus 'drought dodgers', and their special adaptation lies in their ability to complete their life cycle so rapidly.

Succulents — These are perennial and survive the drought by living on water hoarded up from the previous rains. Their structure is adapted to reduce transpiration to a minimum — the compact form gives a small surface in relation to the volume; leaves are often reduced to spines, and there is a thick, waxy cuticle checking any cuticular transpiration. Much of the water is held imbibed by mucilaginous substances and pentosans within the cells, and respiration follows an unusual pattern allowing opening of the stomata at night instead of during the day. An experiment carried out at the former desert laboratory of the Carnegie Institution at Tucson, Arizona bears eloquent testimony to the efficiency of these plants as water hoarders. A large *Echinocactus*, weighing 37·3 kgm. was dug up and kept unwatered in the laboratory for six years. During this period it remained alive and the total loss in weight was 11 gm. (Ashby, 1933). Growth during the dry period is at a standstill; but in the brief wet spells the stomata open wide, allowing both rapid intake of carbon dioxide and transpiration of the water absorbed by the shallow root system. For a short time photosynthesis, respiration and growth proceed apace, until the ground dries up again and the stomata close. It will be seen that this type of xerophyte never really suffers drought *internally*.

Succulents are best known for the many quaint and grotesque forms belonging to the family Cactaceae, but it is worth noting that a number of other families have some succulent species.

Drought Endurers — Structurally, these plants *appear* to be adapted to restrict transpiration. Common features are leathery leaves with thick cuticles, small cells and numerous stomata sunken or protected by the leaves being rolled or densely hairy, and the plant cells have usually a high osmotic pressure. However, when water is available they transpire rapidly, and no measures are taken to store it. When the drought comes the plants wilt and look dead, but vigorous growth and activity are resumed with the next rains. Here the adaptation is physiological; the ability of the plant to survive drought has little to do with its structure but depends upon some property of the protoplasm which enables it to stand extreme desiccation and yet remain alive. In fact, the vegetative body of the plant is endowed with the same type of resistance that we normally associate with seeds.

We may naturally ask what, then, is the significance of the structural peculiarities, known as **xeromorphic characters**, if they are not the means by which the drought endurers survive the dry seasons. Many of these xeromorphic characters are shared by plants growing in quite different habitats, notably salt marshes, heaths and marshland, but these plants too transpire freely, showing that they do not live in a state of physiological drought, as was formerly supposed. Nor are they xerophytes, for they do not live in dry places and cannot survive drought under experimental conditions. A clue to our understanding of this problem comes from observations that some measure of xeromorphy can be induced in ordinary mesophytic plants (for example, white mustard) by simply growing them under dry conditions. The leaf cells are smaller, there are more stomata per unit area, and the cuticle is thicker than in the controls grown under moist conditions — and, as gardeners well know, the plants are hardier. In twining plants like bindweed or black bryony, grown with restricted water supply, an increase in xeromorphy can be seen in successive leaves up the stem; this has been correlated with their falling turgor through competition for water with the lower leaves. We do not know the full answer, but as a working hypothesis it has been suggested that a limited water supply when the leaf is unfolding results in high osmotic pressure in the cells. This appears to be a cause of smaller cell size, and hence (as the proportion of epidermal cells which become guard cells is constant for the species) more stomata per unit area and a higher transpiration rate. Excessive transpiration, then, may be a direct *result* of dry conditions, and it is only partly compensated by true adaptations like sunken or protected stomata, so that the net rate remains high. It can be seen how species having a high osmotic pressure through other causes than direct drought may also show xeromorphic trends.

These general principles are seen most clearly in desert vegetation, but nevertheless have some application among the plants of temperate climates. Some winter annuals which regularly pass the summer as seeds, not germinating until the early autumn, are in fact behaving like the drought dodgers of the desert in avoiding the driest season of the year. As the prevailing temperatures during their growing period are so low we should hardly expect the rapid growth of desert ephemerals. It is interesting that a number of these plants are also Mediterranean in their distribution. There are a few succulents, like the stonecrops and houseleeks, which thrive on the tops of walls and roofs where they have to face quite severe conditions. There is nothing quite corresponding to the drought endurers of the desert, though doubtless some of our species exhibit these properties in a small degree. It is noticeable how mullein (*Verbascum thapsus*) and rose-bay willow-herb (*Chamaenerion angustifolium*) can grow at the tops of high walls and look none the worse for considerable periods of drought. Many mosses, too, will recover after long periods of desiccation. Studies on the experimental production of xeromorphy are quite easily carried out (Ashby, 1934). An interesting related phenomenon is the transpiration check achieved by movement of the leaflets in woodsorrel (*Oxalis acetosella*). They may collapse within a few minutes of coming into bright sunlight or on exposure to a drying wind, while the leaflets on sheltered plants remain erect.

Hydrophytes — Flowering plants represent the climax of a series of evolutionary changes towards adaptation to land conditions, yet, as with insects and mammals, a number of species have gone back to the water environment to exploit its opportunities with great efficiency. This has resulted in profound modifications in anatomy and gross morphology which may be reviewed here, though for details reference should be made to the appropriate textbooks (Arber (1920); Fritsch and Salisbury, 1946). The changes are best considered in relation to the problems (or in some cases the disappearance of problems) which life in the water environment involves.

The need for **mechanical support**, so pressing for land plants, largely disappears in water. The plant tissues, almost invariably rich in intercellular air spaces are very buoyant. While true xerophytes are characterized by much woody tissue, hydrophytes show little, if any, lignification. Xylem vessels often remain feebly lignified, but in species showing more extreme adaptation to the water these vessels disintegrate, and conduction is carried on through canals formed by the spaces which they occupied. Rather surprisingly, a transpiration stream persists, even in totally submerged aquatics. Such mechanical tissue as is present tends to be arranged in a central core, like that of roots, fitting the stems or petioles to take tension rather than compression and bending strains, hence their flexibility. Only in stalks holding the flowers above water level (water violet, arrowhead or water plantain) is the cylindrical arrangement preserved.

The water provides a very different medium from air so far as **gas exchange** is concerned. Carbon dioxide is normally present in much higher concentrations than in the air, but oxygen with its poor solubility becomes a scarce commodity. Also the rival demands for oxygen from animal life and bacteria rotting down organic matter must be very high indeed in habitats like ponds or overgrown canals, which lack the turbulence of swiftly flowing streams to enrich the water with oxygen. Lack of development of cuticle in hydrophytes allows gas exchange (in solution) to take place over the whole surface of the plant, and the submerged dissected leaves common to many species give some increase in the surface/volume ratio. Whether this makes any significant difference seems rather doubtful, and it is reasonably well established that the *cause* of this dissected leaf form is the lower nutritional level of the submerged leaves. Species of water crowfoot have been induced to form their aerial type of leaf under water, given sufficient light and carbon dioxide, while in damp air, free of carbon dioxide they formed dissected leaves. The most effective adaptation to combat oxygen shortage lies in the continuous air spaces in the tissues, which allow diffusion of oxygen released in photosynthesis to all organs of the plant. Floating leaves such as those of the water lilies, frogbit or *Potamogeton* spp. have all their stomata on the upper surface.

Perennation poses problems only to those species with floating or aerial leaves, for below the surface layers the winter temperature of the water will remain at 4° C. (corresponding to its maximum density). In submerged species the buds carrying on the next season's growth are unprotected, turion buds, differing little from the normal bud of the growing season. Species with floating or aerial leaves could not survive the ice: they die back and perennate usually by means of rhizomes (water lilies) or tubers (arrowhead). The behaviour of the free-floating duckweeds is particularly interesting, for in late autumn they sink to the bottom and spend the winter there, rising up to the surface again about the following May. From casual observations it seems that quite a proportion of the plants fail to come up from the bottom in the spring and may die. Investigations of the mechanism of this (*Blundell's Science Magazine*, No. 7), suggest that an increase in the content of stored starch in the fronds may play a part in the autumn sinking, while bubbles of carbon dioxide collecting in the intercellular spaces are probably responsible for the plants rising to the surface in the spring. At that time of year the bacterial decay of last season's dead vegetation has saturated the water with carbon dioxide, so that when respiration in the duckweed fronds at the bottom is quickened by the rising temperature the resulting carbon dioxide appears as bubbles in the intercellular spaces rather than diffusing away into the water. The free-floating rosettes of frogbit (*Hydrocharis morsus-ranae*) have to face a similar problem. In this case, specialized turion buds are formed at the ends of lateral branches, and these become detached and sink to the bottom in the autumn, while the main plant dies.

6

Biotic Factors in the Environment

In the preceding four chapters we have surveyed the impact of the physical environment on the shoot and root system of the *individual* plant. This is, of course, a simplification of the problem, for a plant growing under natural conditions cannot properly be considered in isolation; it must be seen as a member of a complex system of plant and animal populations, all interacting with each other, and usually in a state of delicate balance so that quite small changes may gradually alter the whole system. Herein lies the very essence of ecology. But the simplification is a necessary preliminary to the study of plant communities in Part III, for only when we have become familiar with the part that each factor plays can we hope to appreciate the architecture of the whole. For the time being, then, we shall still focus our attention on the individual plant; the effects of other plants and animals upon it will be regarded as the biotic factors of the environment, while the impact of our individual plant upon the other populations will best be discussed in Section 2 as an aspect of aggression.

Biotic factors may conveniently be classified under three headings:

(1) Effects of man.
(2) Effects of other animals.
(3) Effects of other plants.

To the ecologist man is just another animal species operating in the system of living things, but his influence on vegetation has been so wholesale and so widespread that he does, perhaps, merit a section to himself. Even so, some of the most profound effects arise indirectly, through the changes he brings about in populations of different animal species, and no clear dividing line can be drawn between sections (1) and (2). By contrast, the effects of other plants operate mostly through changes in microclimate which they bring about.

(1) EFFECTS OF MAN

8,000 years ago (geologically speaking, very recently) man had made hardly any impression on vegetation, for he was then a relatively scarce animal, gaining his livelihood by hunting. The great glaciers of the Ice Age had finally retreated from Britain and pine and birch forests flourished in a climate growing rather warmer than we have now. During the following 3,000 years the pine and birch gradually gave place to oak forests over much of the country and Neolothic peoples had come to replace the earlier hunters. They were farmers, clearing the land to graze their flocks, and practising a crude form of agriculture. With them began the destruction of forests and modification of our natural vegetation which has been carried on ever since. At first only upland regions such as the wolds and chalk downs were affected, leaving dense woodland (*horrida silva*) still a prominent feature of the countryside in Roman times; but the changes gained steadily in momentum with the needs of the growing population. It is beyond our scope to trace them in detail here, this has been admirably done by Stamp (1955). Suffice it to say that there is little really 'natural' vegetation, not directly affected by man, remaining in the whole of the British Isles. This need not dismay the plant ecologist unduly; he must simply take account of the intensity and ways in which the human factor is affecting the environment of any particular habitat. In many areas habitats such as sea cliffs, sand dunes or mountain screes have suffered little or no human interference, and there is still much 'semi-natural' woodland, where planting and natural regeneration have left much the same species as those that flourished there long ago. Where the human factor operates more intensely, as in coppiced woodland, hedgerows, grazed meadows or ornamental lawns, there is still plenty of room for adaptation and selection among the plant populations, and even the highly artificial communities of farm crops offer interesting ecological problems especially in relation to the weed flora. Despite man's efforts in cultivation and the use of selective weed killers, the weeds are still numerous, and their rapid return on lawns (presumably from seed) after treatment to remove them can provide a constantly recurring problem for study in schools.

Man's most obvious part in the conversion of natural vegetation to agricultural land lies in the steady destruction and clearing of forests by felling or fire; but the effects of grazing by his flocks and herds are scarcely less important. Particularly in early times large areas of woodland must gradually have perished through the common practice of pasturing animals on the grass growing under the trees, for under such conditions too few seedlings would survive to replace the older trees as they died out. For the same reason regeneration of woodland on land not actually under cultivation is effectively checked by grazing. Thus we see that in the long run grazing may be regarded as a weapon in the hands of man no less potent than the woodman's axe or the bulldozer. Nevertheless, it is obviously to be classified under 'effects of other animals', even if artificially intensified by man, and will be considered in greater detail in Section 2.

There are many less obvious, but still far-reaching results of man's activities. Clearing of the trees paves the way to an increase in the population of wild grazing animals which are normally scarce or absent from woodland habitats — in Britain especially rabbits. The rabbit is not truly a native of the British Isles, but was brought here by man, probably in Norman times, though it does not appear to have shown any spectacular increase at first. Probably most of the early grazing changes were the result of sheep farming, which reached a peak in the fourteenth and fifteenth centuries as witnessed by some of the magnificent 'wool churches', built by wealthy wool merchants, like that at Northleach in the Cotswolds. But by the time that sheep farming declined, there were plenty of rabbits to 'take over' and check the regeneration of trees on the springy downland turf.

Any upset in the natural balance of animal populations may have its repercussions on the vegetation. In woodlands, mice and voles, with some rabbits, take a heavy toll of the tree seedlings; but if the populations are kept in check by their natural predators, chiefly weasels and stoats, enough seedlings may survive to maintain regeneration. If the woods are preserved for game, the check on the population of carnivores imposed by zealous game-keepers may lead to a considerable increase in the numbers of small rodents, with a consequent threat to tree regeneration. On a larger scale this problem has given rise to serious crises in forestry in the

United States, where protection of the wild deer through interference with their natural predators (for example, coyotes and the American puma or mountain lion) has led to over-population of the deer ranges, with dramatic effects on the vegetation. The red deer of our deer forests, formerly a woodland animal of widespread occurrence in Britain, now survives only on upland moors, where its numbers are limited by available food supplies rather than by natural predators.

Another important effect of clearing the forests is the permanent impoverishment of upland soils which it has caused. Evidence from early place-names, from the discovery of the shells of woodland snails, and particularly from the pollen and the remains of big trees buried in peat deposits, all contribute to our knowledge of the distribution of the early forests. It seems clear that these forests clothed not only the downs (where the soil could still support trees), but the mountains of the Pennines, the Lake District and the Scottish Highlands up to at least 2,000 ft. and even 3,000 ft. in some sheltered places. This is far above the tree line of today, and includes wide areas of moorland where the soil is too acid or waterlogged to allow any tree growth. The changes are the result of intensified leaching after the removal of the tree-cover, as explained in Chapter 3 (p. 57).

Among the many other indirect effects of man on vegetation are the specialized habitats which he inadvertently provides for colonization, where peculiarities of the environment may lead to selection of plant populations different from those found under more natural conditions. These can give scope for interesting and worthwhile ecological studies in the most unpromising neighbourhoods. They may not necessarily illustrate the operation of the human factor except in that man 'provided' the habitat, but they merit some discussion here, if for no other reason than to emphasize that the town-dweller can usually find some ecological problems on his doorstep if he really looks for them.

A comparison of the environment and flora of different walls can prove a most interesting study. The mortar and brick or different kinds of building stone provide materials of differing pH and hardness for the colonizing plants. Conditions of light, temperature and exposure to desiccation will vary with the direction in which the wall faces and the degree of shelter, while embryo soils may build up on any ledges which are available. It may be possible to find the age of the wall, or how long it has been left free from human interference, which will give useful information about the rate of growth of lichens on it. For examples of work of this kind see Rishbeth (1948), (*Blundell's Science Magazine*, No. 6). Similar studies can be made on the flora of roofs and the rain gutters of buildings (*Blundell's Science Magazine*, No. 2). In all these cases the source of the colonizing material and its means of dispersal are of special interest.

The streets of cities and large towns provide another kind of habitat in which light-intensity, exposure to desiccation and wind, soil depth and nature may vary to a surprising extent with the direction, slope and camber of the road, and the type of pavement. The correlation between these factors and the numbers and species of colonizing plants gives very interesting material for study, and some detailed results of work along these lines done in Bristol are given by Bracher (1937). Waste places and their flora provide similar ecological problems, with scope for a good deal of individual variation, for example, dockyards, slag heaps, coal tips, etc. One particular case; the colonization of bombed sites in the Second World War received much attention (Fitter, 1945) but interest centred mainly around the *sources* of the colonizing plants and their methods of dispersal, which is an aspect belonging more properly to Part III of this book. In the same way, man's introductions, either as escapes from cultivation or accidental weeds brought into the country with imported merchandise, are more appropriately discussed under 'dispersal'. Introductions of animals are another direct consequence of human interference. Examples with far-reaching effects are the introduction of the rabbit, the Colorado beetle, and in America the Japanese beetle, which has far more obvious effect on vegetation in general because of the wide range of plant species which it will attack. Gardens around New York are devastated year after year by this pest, which occurs in such great numbers that millions of the beetles may be found drowned along the edge of the sea. Considerations of this kind must be given careful thought before the balance of food chains is upset by the introduction of new predators to achieve **biological control** of some troublesome weed or insect pest of crops. A well-known example of successful biological control is that of the prickly pear, a cactus that had over-run wide areas of grazing land in Australia. The introduction from Argentina of a moth, *Cactoblastis cactorum*, the caterpillars of which feed on the prickly pear resulted in the clearing of some 22,000,000 acres within a few years (Wilson, 1950). Every care had been taken before the introduction of the moth to ensure that, when the caterpillars had checked the prickly pear, they would not run amock and turn to other food plants which might be of value to man.

(2) EFFECTS OF OTHER ANIMALS

Obviously, **grazing** is the effect of primary importance. If sufficiently intense it will eliminate not only the tree seedlings, but also many undershrubs and herbaceous plants. Thus, grazing has a selective effect in favouring the survival of those species which are better equipped to stand up to it. But before discussing this we must examine the 'grazing factor' itself in more detail, for it varies with different animals.

(*a*) *Closeness of grazing.* This will play an important part in its influence on the balance of competition between different species. As is well known, cattle do not actually nibble the grass, but twist their tongue around it, then gripping it against the horny pad of the upper jaw (there are no upper incisor teeth) they tear it off with a jerk of the head. As a result, they do not graze nearly so closely as sheep or rabbits, which nibble the plants with their incisor teeth. Badgers, squirrels and the smaller rodents like mice and voles, although equipped to nibble closely, rely on a more varied diet, and cannot really be regarded as grazing animals. They may nevertheless have a considerable effect on vegetation, check-ing tree regeneration by eating the seedlings or seeds and nuts.

(*b*) *Selection of species attacked.* As a rule grazing animals prefer the more succulent species and tend to avoid plants which are more fibrous, hairy or prickly. Preferences, of course, vary to some extent with different animals; horses and sheep are more fastidious than rabbits or cattle, and it is traditional that donkeys will eat thistles, and goats almost any plant. But as goats and donkeys rarely form an important element in our grazing fauna, it remains generally true that the coarser vegetation is less harmed by grazing. A number of species such as elder, bracken and ragwort enjoy immunity from animal attack because their shoots contain substances that are either distasteful or actively poisonous.

The lawn mower may conveniently be considered as a grazing animal which nibbles closely but exerts no selective choice at all, though wiry grass and plantain inflorescences may resist its attentions to some degree if it is in need of sharpening. As a non-selective 'animal' it could prove useful as a control in experiments designed to show the effects of selective grazing.

(c) *Effects of trampling and burrowing.* It is well to remember that trampling, with its inevitable damage to plant shoots, as well as consolidation and often puddling of the surface layers of the soil, is a factor which must always accompany grazing. That this can have profound effects on the species making up the vegetation was shown most convincingly by Bates (1935) in his classic study of the flora of footpaths. Even when the grazing animals are rabbits the damage from treading is by no means negligible, as may be seen from the absence of unpalatable species like bracken around the burrows. But here the effects of burrowing are important, for the loosening of the soil leads to erosion and drying out of the surface layers which favours the survival of the more deeply rooting species.

We are now in a better position to examine the selective effects of grazing on plant communities. As a matter of convenience, the plants concerned may be roughly divided into three groups:

(1) Avoided species.
(2) Resistant species.
(3) Non-resistant but palatable species.

Avoided species. These will obviously be at a great advantage where the grazing is intense. Thus elder (*Sambucus nigra*) often forms flourishing communities around rabbit warrens or badger sets of long standing, where it can grow largely free from competition from other plants. Indeed, a conspicuous stand of elder may often be the first means of locating a rabbit warren or badger set (Figs. 6.1 and 6.2). Smaller plants like forget-me-nots (*Myosotis* spp.) and ground ivy (*Glechoma hederacea*), which are avoided probably because of their hairiness, can also be found growing close to the burrows. Two other avoided species, ragwort (*Senecio jacobaea*) and creeping thistle (*Cirsium arvense*) are notorious weeds

FIG. 6.1. Chalk down capped with elder thicket covering extensive rabbit warren. Beech tree on right gives the scale. Near Folkestone

FIG. 6.2. Elder thicket with badger set, Cheesefoothead, near Winchester

of overgrazed pastures, where the rival plants in the sward are too much weakened by grazing pressure to compete successfully against them.

But the selective effects of grazing preference are usually much more subtle than this, for the animals, particularly rabbits, show **degrees of preference** for different species, which can lead to a distinct zonation of vegetation around the burrows. Taking a rather simplified illustration from heathland plants near the sea; rabbits like tree seedlings best, eating shoots of ling (*Calluna vulgaris*) and sand sedge (*Carex arenaria*) only as second or third preferences. If the rabbit population is very small, the supply of tree seedlings may largely meet their needs and ling can grow near the burrows, but any increase in their numbers will mean that they will have to eat ling and other species which they like less. Bracken is avoided and is thereby helped in its competition with the ling. As the rabbits will not go further than they need for food, the grazing pressure is most intense nearest to the burrow, and a zonation, like that shown in Fig. 6.3 develops. Elder would not be present around the burrows in this case as it is intolerant of the acid soils characteristic of heathland. If the rabbits are excluded, or move away, rapid changes in the vegetation result.

Resistant species. Resistance to rabbit attack is largely a matter of growth form. The obvious essentials are that the food storage organs and the buds for making fresh growth should be adequately protected, and it is therefore to be expected that the majority of resistant species will be either hemicryptophytes or geophytes (see p. 92). Actually, the hemicryptophytes provide most of the resistant species, perhaps because the aerial portions of geophytes are commonly more conspicuous. Rabbits undoubtedly go for the taller, more conspicuous plants, and some of the hemicryptophyte rosette forms may lie so flat on the ground as to escape grazing altogether. At the same time they are able to cut off the light from any low-growing plants around them, and can thus put up very effective competition. The rosette forms like plantains and daisies, so common as weeds in lawns provide familiar examples which find their natural home in closely grazed swards.

Grasses enjoy a further advantage over dicotyledonous species in that their leaf-blades grow from a meristematic region at the base which continues active for a long time. If therefore the end of the leaf-blade is nibbled off, the effect is to stimulate further growth rather than to kill the leaf. But it still remains true that those grasses which spread by stolons have their buds vulnerable to attack and are not able to stand up to intense grazing.

Another aspect of resistance lies in the ability of a plant species to modify its habit of growth when damaged by grazing. The removal of the growing point will naturally stimulate the development of dormant axillary buds, and this may lead to changes in

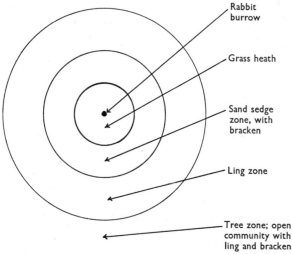

FIG. 6.3. Diagram of zoning of vegetation round a rabbit burrow

the habit of growth of the plant which make it much more resistant to further grazing. Gimingham (1949) has made a detailed study of how this affects competition between bell heather (*Erica cinerea*) and ling (*Calluna vulgaris*). In areas where these two species are normally growing together he showed that on ungrazed hillsides the balance of competition is in favour of the bell heather, but moderate grazing shifts this balance in favour of ling. This, he showed was due to the different responses of the two species to removal of the main growing axes. While the heather produces a number of new erect branches, vulnerable to further animal attack, ling tends to assume a spreading form, with branches growing close to the ground. These are less liable to damage from further grazing and, at the same time, help the plant in competition with other species. Under heavy grazing ling loses its competitive advantage, as both species assume a compact, bushy form.

Non-resistant, palatable species. Little need be said of this group beyond calling attention to the fact that the terms are comparative; there are degrees of 'non-resistance' and palatability. As we have seen, it is the tall, conspicuous plants and the juiciest ones that are most liable to attack, but in times of food shortage the grazing will be less selective and the animals will turn to coarser species.

It is obvious from what has been said that grazing will have profound effects on the vegetation of any area: some species will be eliminated altogether, except in places like cliffs or quarries inaccessible to the grazing animals; others will be greatly weakened thus favouring the increase of the resistant or avoided species in their competition with them. Contrary to what one would expect, the result is generally a *greater* variety in the flora, for a number of species (sometimes rare ones) are able to maintain themselves where otherwise they would be swamped by a dominant grass, were it not kept in check by the grazing. This selective effect makes a fascinating ecological problem which has received much detailed study. A particularly interesting comparison has been made between the vegetation of two small islands off the Pembroke-shire coast; Skokholm, which has supported a vigorous rabbit population for hundreds of years, and Grassholm which has remained entirely free from grazing mammals (Gillham, 1953–6). Though conditions of soil, climate and exposure on these two islands are comparable, their vegetation is entirely different (Figs. 6.4 and 6.5). Grassholm is largely covered by a tall growth of creeping fescue grass (*Festuca rubra*), a species non-resistant to rabbits and consequently scarce on Skokholm where the main communities are bracken with stunted ling in the more sheltered situations, grading to grassland dominated by York-shire fog (*Holcus lanatus*) and a close sward of thrift (*Armeria maritima*) with buck's horn plantain (*Plantago coronopus*) on the more exposed side. As Dr. Gillham points out, Yorkshire fog is a species normally avoided by rabbits because of its hairiness, and owes it dominance on Skokholm to the grazing factor. However, the grazing there is so intense that it is by no means immune, and the swards are kept down to about one-third the height of the Yorkshire fog growing on Grassholm. The species is able to adopt

FIG. 6.4. Skokholm vegetation. (See text)

FIG. 6.5. Grassholm vegetation. (See text)

a resistant habit and can therefore survive grazing quite well. The thrift and buck's-horn plantain growing in the more exposed places are very much modified in their habit of growth as a result of grazing and wind action. The thrift occurs as dense, smooth cushions and the plantain as small rosettes, closely pressed to the ground.

The effects of trampling and burrowing, which must inevitably accompany this rabbit grazing have also been analysed for the islands (Gillham, 1956). As might be expected, the concentrated treading was almost invariably found to have adverse effects on the vegetation. The manuring which goes with it tends to make plants more vulnerable, as it encourages rapid, succulent growth which is more easily damaged. Avoided species which are immune to actual grazing, such as bracken, may be completely eliminated where the traffic is particularly heavy, for example, around a warren. Treading also has a selective effect in encouraging prostrate and rosette species, mostly hemicryptophytes. It is interesting that in some of the sea-bird colonies, autumn and winter annuals are much in evidence, taking advantage of the period from late summer when the birds have left and the ground remains undisturbed.

Whilst the Pembrokeshire islands provide an ideal opportunity, beyond the reach of most of us, there are plenty of chances to study grazing and trampling effects in less spectacular conditions. Comparison may be made between grazed vegetation and that growing in places inaccessible to animals, though, in the case of rabbits, such places are often not very accessible to the student. The changes resulting from myxomatosis killing off the rabbit population are likely to be far-reaching and are well worth careful study. Finally there is the experimental technique of excluding grazing animals from a small sample plot by fencing or wire netting. To keep out rabbits the wire should be buried to at least a foot, and turned slightly outwards at the base. Mice will climb the wire and it may be necessary to roof over the enclosure with wire netting if one wishes to be sure of excluding them. This will interfere with the light reaching the vegetation, so that control plots used for comparison should really be roofed with wire netting, but have no sides. It is worth while remembering that in the fencing or wire protecting young trees in parkland an experiment of this kind is unwittingly set up, and the results may be already there waiting for study. Even the familiar hen-run provides a special case of an experiment on the effects of trampling and disturbance of the soil by the birds' scratching.

The differences between the vegetation of the experimental plot from which animals have been excluded and the control outside may not be immediately apparent, and a little thought is needed in choosing the best technique by which a comparison can be made. Purely subjective impressions are often very unreliable, and lists of species present or absent give only limited information. More precise data can be obtained from the detailed study of a number of small sample plots or **quadrats**. From these such measures as the frequency, density or percentage cover by different species are determined, and accurate comparisons can be made between the experimental plots and the control. Further details on sampling and the use of these different measures are given in Chapter 11. Where zonation of vegetation is being studied, as round a rabbit burrow, **local frequency** provides a useful measure. For this we need a quadrat frame subdivided into smaller squares by pieces of stretched string (like that used for mapping root systems, p. 31). If a metre quadrat divided into a hundred squares of 10 cm. side proves too cumbersome, smaller units can be used, for example, a 25 cm. square, divided into 25 squares of 5 cm. side. The percentage local frequency of the species in question is obtained for each position of the quadrat by simply counting the number of squares in which it is present. In collecting information of this kind the quadrats are best placed according to a grid system, so that the results can be mapped. The first step is to adopt a suitable base line between two convenient landmarks such as trees or conspicuous bushes; then at intervals of, say, 6 ft. or 10 ft. along the base line a measure or surveyor's chain can be laid at right angles, and quadrat counts made at definite intervals along this. Some sort of picture of the distribution can then be built up by drawing lines to separate off, for example, places where local frequencies are less than 15 per cent; between 16 and 30 per cent; between 31 and 45 per cent and so on. The principle, with hypothetical results, is illustrated in Fig. 6.6.

Measures of frequency, density or percentage cover do not give any *direct* information about how the vigour of growth of the

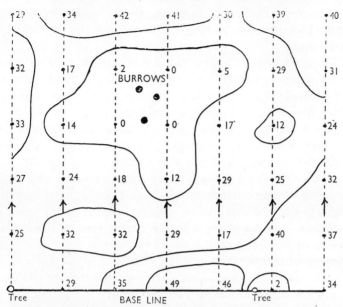

FIG. 6.6. Hypothetical results illustrating the use of local frequency determinations. Forty-one sample quadrats are laid out in a grid to examine the zonation of one species round a rabbit warren. The figures represent percentage local frequency recorded in each quadrat. Quadrat size must be large in relation to the size of the species sampled

species concerned may be affected by grazing. A decline in frequency or density of one species in favour of another will, of course, suggest that its vigour of growth is decreasing, but if competing species are equally affected by grazing no such shift of balance may result. Descriptive notes about the height and habit of growth will provide some of the required information, but sometimes it is necessary to harvest the material from sample quadrats and compare the dry-weights of the yield.

(3) EFFECTS OF OTHER PLANTS

The effects of other plants, though the most important of the biotic factors, fall largely outside the scope of this chapter, for they are mostly indirect, working through changes in microclimate such as light and shade, shelter, etc. These have already been discussed in Chapters 4 and 5, while Chapters 7–9 deal with the means by which vegetative spread, reproduction and dispersal are achieved. There are, however, some direct effects which should have mention, as they can sometimes assume ecological significance.

Fungus diseases of higher plants immediately come to mind. Under natural conditions the host plants will usually be scattered, so that only by a prodigal output of spores and very good dispersal is the fungal parasite likely to maintain itself. Also a natural balance is usually established between parasite and host whereby the host survives the onslaught, for its continued wholesale destruction can only lead to the extinction of the parasite. The scenes of widespread devastation by fungal parasites are usually the result of man's interference. He may unwittingly carry the fungus to a new country where a susceptible host plant grows in abundance — as happened with the blister rust of white pine, introduced into America in 1906. More commonly, by assembling a pure stand of host plants as an agricultural crop, he provides the parasitic fungus with such a golden opportunity that the efficiency of its spore dispersal becomes all too obvious — the epidemic of coffee rust which ruined the industry in Ceylon in 1869 is a classic example.

Reference has already been made in Chapter 3 to the part that antibiotics, secreted by soil micro-organisms, may play in changing the balance of soil populations. Where this affects the mycorrhizal fungi growing in association with forest trees, it could have far-reaching results in determining the success or otherwise of tree growth.

Finally, so far as flowering plant parasites, such as dodder or mistletoe are concerned, the effects on their hosts are so limited that they can have little part in changing the ecological balance of any plant community.

PART II AGGRESSION

7

Shade cast: Vegetative Reproduction and Spread

Nowhere in Nature's grim democracy is the struggle for existence between living creatures sterner than among plant communities. To realize this one has only to pause and consider the staggering rate of increase that would result if *all* the seeds produced by any species of plant were to germinate and come to fruition. An average specimen of groundsel, for example, may easily bear some 1,000 seeds (dispersed as fruits), and these plants normally pass through at least two generations in one season. At this rate of increase, a single plant could produce in three years enough progeny to populate an area more than twice the size of the British Isles, spacing the plants only 2 in. apart. Nor is this an exceptionally high reproductive capacity, indeed, 1,000 seeds per plant is a modest output compared with many far less common species. Yet, as Darwin pointed out, the populations of different species tend to remain more or less constant, so the mortality-rate must be tremendous, and one of the most important factors responsible is competition from other plants. With such large numbers as these, very small changes in the percentage mortality can lead to differential survival rates profoundly affecting the balance of populations of different species.

We have already seen that competition between plants is, above all else, for light, and that the evolution of upright stems in land plants might be interpreted as a response to the biological advantage of being able to form leaves at a level where they will not be overshadowed by other competitors. But this 'self defence' becomes 'aggression' for the smaller plants growing below, especially if it is accompanied by lateral spread; and in the growing forest we have, in effect, an 'armaments race' being carried on; of course, entirely in the interests of self defence! Aggression is also a by-product of making good use of the light, for the more efficient a plant is in this respect, the denser the shade it will cast, making it a more formidable enemy to the lesser plants beneath it. So our next task is to examine the factors which contribute to the aggressiveness of different species, and see how they can affect the balance of plant communities.

HEIGHT

Height is the first requirement if a plant is to be aggressive; so that it can overtop its neighbours and cut off their light. This is, of course, purely relatively speaking, for the effect is achieved equally well by rosette plants growing on lawns or closely grazed turf, but usually some considerable growth of stem is involved. This requires the diversion of food material to provide the necessary mechanical support, and the rate of growth of the plant as a whole must inevitably be slowed down in providing it. Such plants, especially trees must pass through a vulnerable stage before they become established. Once through this critical stage they become the dominant features of the vegetation in that they determine the light-climate beneath them, and hence govern the type of plant which can grow there. Tall, perennial herbs must pass through this stage each year; but for them it is less critical as they have generous food reserves below the ground, and if one shoot is damaged (for example, by grazing) another bud can develop to replace it.

Where the support of taller plants is available, **climbers** take advantage of the opportunity provided and can achieve remarkably rapid growth as so little material need be diverted to the formation of mechanical tissue. It is the conducting tissue which is most prominent in the anatomy of their stems, since the area of cross-section is usually small in relation to the surface of the leaves which they carry. The bicollateral bundles of marrows and cucumbers, with their giant xylem vessels and double set of wide sieve-tubes, afford a good example.

Except for the dodders (*Cuscuta* spp.), which have scarcely any chlorophyll and lose all contact with the ground, climbers in Britain take no food from the plants which support them. They are however aggressive in that their leaves must cut down the

light available, thus weakening the supporting plants, though they
are rarely killed. In temperate climates climbers are most in
evidence in scrub vegetation and hedgerows, where some species,
such as goose-grass (*Galium aparine*) must play an important part
in reducing the light reaching the ground vegetation. The highly
specialized adaptations by which these plants are enabled to climb
are well worth careful observation and study, though an account
of them is beyond our scope here.

LEAF MOSAIC

Given adequate height in a plant, our next concern is how
efficiently it can intercept the light. We know that this can vary
a great deal between different species (one has only to compare a
beech and a birch tree), and it will clearly depend to a large
extent on how good a leaf mosaic the plant can form. This is
governed by characteristics which are a definite part of the make-
up of the species — phyllotaxis, leaf-shape and the phototropic
responses of the young growing stems and leaf petioles.

The general symmetry of an upright stem in normal light is
radial, with the leaves either opposite and decussate in pairs or
arranged in a spiral. The angle between the insertion of successive
leaves when they are spirally arranged commonly approximates to
137°. The causes underlying this regular arrangement of the
leaves on the stem are doubtless to be sought in such factors as
competition for space and food materials in the earliest stages of
development in the bud or diffusion gradients of inhibitory
auxins from older leaf primordia. As a result, successive leaf
primordia arise so far as possible from each other in the apical
meristem, and the mature leaves are spread so that there is a
minimum of mutual shading.

While this gives the ideal arrangement for an upright stem in
the open, it will not meet the case of a creeping stem or a horizontal branch of a tree, nor yet an upright stem illuminated from the
side, as occurs in a hedge. Here, although the basic phyllotaxis
remains unaltered, the leaves as they develop turn to face the
incident light, giving the shoot isobilateral symmetry. The
mechanism is a phototropic response of the petioles or sometimes
a twisting of the stem itself. It is in some cases so nicely adjusted
that the leaves of a horizontal branch become arranged in one

E

plane, into which they fit themselves almost perfectly, with
scarcely any overlap nor any wasted space — a literal mosaic.
There can be no better example of this than the branches of a
beech tree; by contrast a birch or an ash forms a poor mosaic and
neither casts a dense shade. Among herbaceous plants the leaf
mosaic effect may vary with the light conditions in which they
themselves are growing. This is well illustrated by the different
arrangement of the pinnæ of bracken fronds growing in the open
and in shade. In the shade the general trend of growth remains

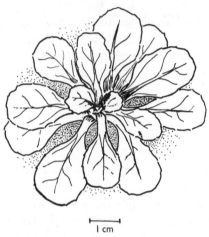

I cm

FIG. 7.1. Rosette of common daisy. Both leaf shape and arrangement
contribute towards avoiding mutual shading

upwards and there is little lateral spread of the upper part of the
fronds, so that no continuous canopy is formed comparable with
that of bracken growing in the open.

The arrangement of the crowded leaves of rosette plants
presents particular problems, if mutual shading is to be avoided
and also emphasizes the importance of the part that growth form
may play in competition. They are usually borne in spirals with a
phyllotaxis ratio high in the series, that is, a large number of
leaves is formed before one occurs immediately above the starting
point. The actual leaves are commonly more or less wedge-shaped
so that they fit together like so many pieces of a cake, while the
younger leaves, being shorter, will shade only the petioles or the
narrow basal portions of the blades of the mature leaves (Fig. 7.1).

Some species with opposite and decussate leaves have assumed the rosette habit of growth, and it is interesting that these may also in fact achieve a spiral arrangement of their leaves (instead of four vertical tiers) by a slight twisting of the stem, like that seen in twining plants. Raunkiaer cites *Crassula orbicularis* as a particularly good example.

While these devices achieve optimum use of the available light, they will make their possessors formidable competitors as they will exert maximum shading upon any plants growing beneath them. This is particularly the case on lawns or closely grazed swards, for under conditions of optimum light the leaves of rosette plants grow out flat and close to the ground, indeed they may actually press downwards by differential growth of the petioles. In long grass, however, the response of the leaves of the rosette plants is changed by the partial shading, so that they grow upwards and do not overshadow their neighbours (Fig. 7.2). The bearing of this on the aggressiveness of rosette plants is well illustrated by simple experiments that can be carried out on any lawn. If part is kept closely cut, and another part left uncut as a control, an increase in the population of rosette plants on the short grass will soon become apparent. Two factors are here at work: the increased aggressiveness of rosette plants when they spread their leaves flat in bright light, and the weakening of the competing power of the grass by regular cutting. Further evidence about their relative importance might be gained from a second experiment in which the rosette plants are made to abandon their spreading habit by shading with a frame of wooden slats, but the grass around them is still kept short by shearing.

SEASON AND DURATION OF SHADING

This is of great importance, as we have already seen when discussing the influence of different tree species on the ground vegetation. A similar kind of relationship can hold good between herbaceous plants; the larger ones, if they are late in their spring growth, leave an opportunity for any low-growing herbs, such as ground ivy (*Glechoma hederacea*), which can take advantage of this early spring light-phase. Also by virtue of this, many winter annuals, though characteristically plants of bare ground, may linger on in hedge-banks which support much taller vegetation

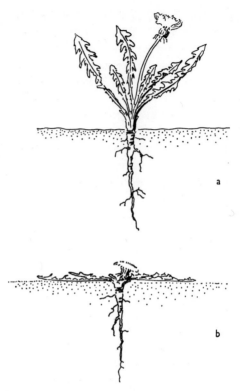

FIG. 7.2. Rosette of dandelion grown (*a*) in partial shade and (*b*) in full light

throughout the summer; the rosettes of hairy bittercress (*Cardamine hirsuta*) are a good example. Under conditions of moderate woodland shade bluebells and bracken are enabled to live this 'Box and Cox' existence, for by the time the tall bracken fronds cast their heavy shade in June the bluebells have already completed their period of photosynthesis and their leaves are dying back. However, where better illumination allows a more vigorous growth of bracken the plants become much more aggressive since the dead fronds, which are very slow in rotting down, gradually accumulate to form an effective barrier to light reaching the bluebells in spring. Less than 10 per cent of the available light may penetrate to ground-level, and in these conditions the bluebells will die out. This resistance to decay of the dead bracken fronds virtually enables the plant to behave as though it were an ever-green, so far as aggression is concerned. The dead leaves of many conifers are also very slow to decay, because of the resins that they contain, but here the main effect on other vegetation is indirect, in that their accumulation leads to increasing acidity of the soil.

ROOT COMPETITION

The study of plant competition above the ground is made easier by the fact that the part played by light obviously dominates that of all other factors. Below the ground, however, the situation is much more difficult to assess, for one cannot easily disentangle the relative effects of competition for water and for soil nutrients. There can be no doubt that root competition may sometimes be of great importance.

Agricultural data from experiments with different spacings of crop plants give convincing evidence in support of this; for example, in the United States the yield per acre of pecan nuts has been shown to be greater if the spacing between the trees is increased beyond that at which mutual shading is eliminated. In the garden, the poorer growth of cabbages or Brussels sprouts planted too closely is more marked than one would expect from the mutual shading effects alone. Again, in desert vegetation the bushes are always well spaced out, sometimes even with a regu-larity that simulates an orchard. In this latter case the competition is obviously for water, but under average British conditions, where water shortage is rarely a master factor, one suspects that competi-tion for mineral salts may loom more important. A. S. Watt and G. K. Fraser (1933) carried out some interesting experiments in an attempt to gain evidence on this point, studying the growth of wavy hair-grass (*Deschampsia flexuosa*) and wood sorrel (*Oxalis acetosella*) on the floor of a pinewood. They found that cutting deeply around the sides of experimental plots, and so severing the competing pine roots, resulted in a marked increase in the yield (dry weight) of both species. This clearly indicates that competi-tion from the pine roots is effective in checking the growth of the ground flora.

That this competition is primarily for mineral salts is suggested by the results of another experiment where watering the plots (and cutting down the sides only to a depth of 4 in.) gave no beneficial effects at all on the yields obtained. Examination of the root systems showed that the wavy hair-grass had deep roots, in direct competition with those of the pine, but those of the wood sorrel rarely penetrated below the humus layer in the top 2–3 inches of the soil. This makes the increased yield of the wood sorrel hard to understand. There is scope for much experimenting of this kind; beechwoods on chalk should be a profitable subject, where the shallow rooting habit of the trees, and the rapid drainage of the soil suggest that competition for water may be intense. It is important that when investigating any problem of competition between species, the distribution of their root systems in different soil horizons should be examined.

VEGETATIVE SPREAD

The growth in height, heavy shading and root competition discussed so far in this chapter, although essential ingredients of aggressive behaviour in plants, cannot do more in themselves than establish purely local supremacy for the plants possessing them. They must be accompanied by some means of spreading if active aggression is to take place and new territory be 'conquered'. Seed dispersal (to be discussed in Chapter 9) provides for long-distance spread of the species, but has the disadvantage that the resulting new individuals must pass through a highly precarious stage in their existence until they have become established. By contrast, vegetative spread, which we must consider now, is attended by far less risk for the new plants and consequently may often prove a more effective means of aggression, although, of course, the rate of spread is necessarily slow. It is achieved in a variety of different ways, by adaptations in structure and behaviour which seem to be relatively simple. Yet in many cases we do not really understand the underlying mechanism, and there is here a field in which we can discover a great deal from simple experiments on the behaviour of different plants.

A fairly close relationship is usually evident between over-wintering (perennation) of herbaceous plants and vegetative *reproduction* — as opposed to *spread*. When the flowering phase is reached in annual (monocarpic) species, the apical meristems of all the branches go over to flower production, so that there is no longer any replacement of the older leaves as they die off and the plant loses its capacity for manufacturing food. Even though there are still some dormant axillary buds at the base of the stem, there appears to be no food left for their development, and the plant dies exhausted by the crisis of flowering, quite irrespective of the onset of winter. The effects of removing the flower buds as they form are interesting; when this was done to charlock (*Sinapis arvensis*) the natural tendency of the plants to woodiness was checked, while considerable local swelling due to accumulation of starch occurred at the junction of the stem and root, with the development of numerous small branches in that region (*Blundell's Science Magazine*, No. 5).

Might the plant survive the winter to grow and flower again if the number of flowers was restricted by disbudding, so that some food was left to develop the dormant buds at the base of the stem? This is the kind of thing that happens normally in many perennials (polycarpic plants), where the proportion of food used in flower and fruit production is naturally much lower. When the stems die back after flowering the lower parts remain alive, and being rather woody they are resistant to winter cold. Dead leaves or snow may give the dormant axillary buds some protection from exposure, and there is enough food stored underground for their development into new shoots in the following spring. This cycle repeated year after year, gives the tufted growth characteristic of plants like Michaelmas daisies. In time the number of stems has greatly increased, and they eventually become separated into individual plants by decay of the underground organs connecting them. Thus, vegetative reproduction is achieved, but with scarcely any territorial spread. A similar state of affairs is seen in bulbous plants, like bluebells, snowdrops or daffodils, and also the corms of crocuses. In such plants as these there is specialized adaptation for food storage to provide for the development of next year's bud, provision to keep this storage at a more or less constant level in the ground by contractile roots, and a seasonal rhythm in leaf production enabling them to make use of the spring light-phase in woodland, but no real adaptation for vegetative spread. Nevertheless a very gradual spread is achieved which may effectively prevent competition from other taller species by sheer weight of numbers. It has thus an advantage over spread by fast-growing runners or rhizomes which leave open pockets of ground for colonization by rivals, and the resulting plant societies are usually stable. The dense carpets of bluebells illustrate this point, though it must be remembered that regeneration from seed plays some part in their formation.

Special adaptations for vegetative spread mostly take the form of changed tropic responses and rapid growth in some of the lateral branches; the normal more or less upright growth is replaced by branches creeping horizontally along the ground and rooting at the nodes. Of course, lateral branches always occur at an angle to the main stem, and stems in contact with moist

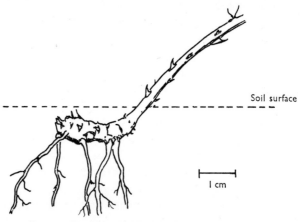

Soil surface

1 cm

FIG. 7.3. Rooting end of a bramble branch (October)

ground will commonly form adventitious roots. The arching over of bramble stems (*Rubus* spp.) is simply the result of their mechanical weakness as scrambling plants, their peculiarity lies in the ease with which they take root and change their direction of growth where the ends of their stems have made contact with the ground (Fig. 7.3). But the creeping stems or **runners*** specially adapted to the function of vegetative spread are not constant in their responses to light and gravity. As a rule they make rapid growth along the ground at first, with long internodes, but eventually turn upwards forming a short, rosette-like rootstock with crowded nodes and numerous adventitious roots. Commonly this becomes swollen with food (as an adaptation for perennation) and gives

---

* The term **stolon** has been avoided as it is so often used with different meanings.

rise to an erect flowering shoot the following year, while buds developing from the axils of the lower leaves grow out to form fresh runners spreading in all directions. The actual runners may bear normal leaves, as in bugle (*Ajuga reptans*), (Fig. 7.4*a*) and differ from the upright shoots only in their tropic responses. But they are often more specialized in that the internodes are abnormally long, for example, creeping buttercup (*Ranunculus repens*), and the leaves along them may be reduced to mere scales, for example, strawberry (*Fragaria vesca*, Fig. 7.4*b*).

Anybody familiar with gardens will know how rapidly these plants can spread — there is no question about the success of this method of aggression. But one would naturally like to know more about the factors governing the behaviour of the runners in different species. Why do they take on a creeping habit of growth in the first case, and why do the ends later turn upwards to form a vertical flowering shoot? What brings about the change from the long internodes formed at the creeping stage to the telescoped stem of the terminal rosette? Doubtless auxins control these changes (and it should be possible to verify this experimentally by application of different concentrations in lanolin), but we are then faced with questions about the *causes* of the varying auxin concentrations in the plant. What, in fact, are the external factors favouring or inhibiting the formation of runners and their elongation?

The Swedish botanist Lidforss showed many years ago that the geotropic response of the stems of plants (for example, *Lamium purpureum*, *Stellaria media*, *Chrysanthemum leucanthemum*, *Veronica hederifolia*) may vary with temperature; they will grow vertically at high temperatures, but horizontally if the temperature is low. Light may also be important: the phototropic response in some plants is known to vary with differing light intensities, or, alternatively, under conditions of heavy shading the shoots may be too feeble to grow upright, as the behaviour of yellow archangel (*Galeobdolon luteum*) suggests. It is noticeable with a number of woodland species (for example, yellow archangel, creeping Jenny, bugle) that shaded, moist habitats favour vegetative spread by runners, while in drier, sunny situations the plants flower readily and produce fewer and shorter runners. Salisbury (1942) gives an interesting comparison of the behaviour of a patch of yellow

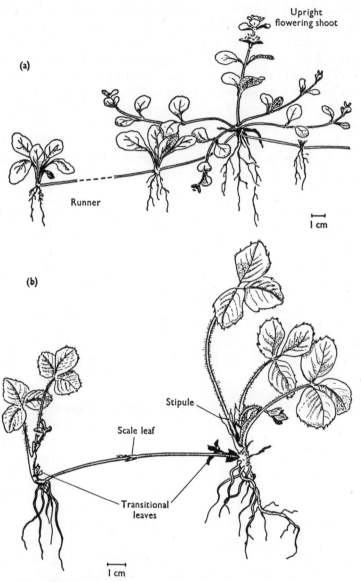

FIG. 7.4. (a) Bugle (*Ajuga reptans*) showing runners and creeping stems with normal leaves. (b) Strawberry (*Fragaria vesca*) — the runners have leaves reduced to scales. (After Priestley and Scott, 1955)

archangel in two successive summers, the first dry and sunny, and the second dull and wet. Situated in partial shade, the plants flowered freely in the dry summer, but produced fewer runners (3–4 per plant) than in the following wet summer (4–5 per plant), when the number of fruits set was about halved. The runners produced in the wet summer were markedly longer, as the following table (Table 7) shows, and they bore a higher number of rooting nodes.

TABLE 7. *Lengths of Runners of* Galeobdolon luteum *produced by the same Plants in a Dry and a Wet Season.*

Class intervals* of runner length (cm.)	Percentage of population in each class	
	Dry summer	Wet summer
5–20	4·3	—
20–35	8·7	—
35–50	21·7	8·3
50–65	33·3	13·3
65–80	18·8	16·6
80–95	11·5	16·6
95–110	1·4	20·0
110–125	—	10·0
125–140	—	11·6
140–155	—	3·3
Average length of runners	59 cm.	91·5 cm.

Light undoubtedly influences elongation of internodes, as is shown by the reaction of a buried dandelion plant, which grows rapidly to the surface and forms a new rosette. But the behaviour of runners would be difficult to explain in terms of varying light intensity alone, even taking into account seasonal shading due to the tree canopy in woodlands.

Another aspect of the light factor where much can be discovered by simple experimentation is the effect of length of day (or, more correctly, length of darkness) not only on flowering but also on the habit of growth. It has been shown that under some conditions variations in length of day will change the geotropic response of chrysanthemum stems so that they will adopt a creeping habit. Some such response may explain the peculiar behaviour of the

* See page 141.

lateral branches of greater bindweed (*Calystegia sepium*) formed from about August onwards. These grow straight along the surface of the ground with the singleness of purpose of a Roman road, and will pass right through a raspberry thicket without any attempt to climb the numerous supports available there.

Experimentally, the length of the daily period of darkness may be increased simply by covering over the plant concerned with a box at a fixed time every evening during the long days of the summer. It is of the utmost importance that the treatment should be strictly regular; forgetting about it only once may vitiate the results of the whole experiment. If the plants being studied can be grown in pots or boxes there is no great difficulty in providing them with extra daylight during short days, and so shortening the daily period of darkness, for the light *intensity* makes little difference: a 40-watt bulb in a garage or greenhouse is quite adequate. The experiments should normally be started as soon as the growing season commences, or the seeds have germinated, as the length of day experienced at these early stages may effect the subsequent behaviour of the plants.

Finally there are internal developmental changes in the plant which may influence its vegetative growth, probably through the medium of auxins. Whatever induces the buds to go over to flower production instead of leaves, be it auxins or anything else, may well have secondary effects on the branching behaviour. The change in the nutritional balance resulting from flowering is bound to influence vegetative growth, and once the terminal stem has flowered and fruited, its auxin production will no longer be a factor exerting some control over the behaviour of the lateral branches.

From the point of view of aggressive vegetative spread there is really little difference between runners above the ground and rhizomes running more or less horizontally beneath its surface. The distinction concerns rather the degree of protection from grazing, frost or drought which the storage organs and resting buds enjoy. As a rule the runners connecting newly formed rosettes with the parent plant die in the winter, thus separating the individuals at an early stage, while an underground rhizome connection generally remains for some years before decaying. For this reason rhizomatous plants often play an important part in stabilizing sand dunes or tidal mud; marram-grass (*Ammophila arenaria*) and rice-grass (*Spartina townsendii*) are widely planted for this purpose. The association between perennation and vegetative spread is perhaps closer in rhizomes than it is with runners, so that they might be regarded as representing a slightly less specialized stage of adaptation. As, usually, buds only form from stem structures it is natural to find modified stems as the commonest storage organs, and if these are situated underground they will be sheltered from exposure to winter weather. As a rule, the allocation of stored food in rhizomes is a generous one. All that is needed then, to accomplish vegetative reproduction is that more than one resting bud should develop, while horizontal growth of the rhizome below the ground will achieve lateral spread.

The growth pattern in rhizomes is generally sympodial as in the case of most plant with runners, that is, at the beginning of each season the terminal bud of the rhizome grows upwards to form the aerial shoot system, while the functions of the rhizome (storage and lateral spread) are carried on by one or more buds borne in the axils of scale-leaves below the ground. In the succeeding season their turn will come to grow up above the ground to produce leaves and flowers. The tropic behaviour of rhizomes is more complex than that of runners in that they can usually maintain themselves at a more or less constant depth in the ground, despite irregularities in the soil surface. On reaching a bank where the ground slopes upwards, they too will grow upwards in the soil; if the rhizome is set near the surface of the soil, it will grow sharply downwards to the required depth. In effect, these plants are able to place their winter buds at a particular depth in the soil. The mechanism of this response is not fully understood. There is evidence that the concentration gradient of carbon dioxide in the soil may play a part in some cases; but Raunkiaer (1907) showed that in the case of Solomon's seal (*Polygonatum multiflorum*) it is a light-response of the aerial stem. By enclosing this in a darkened cylinder, up to about one foot above soil level, he was able to make the rhizome grow upwards in the soil, even though it was already near the surface (Fig. 7.5). In the controls, growing at a depth of about 5 cm. the continuation bud grew horizontally or slightly obliquely downwards. This experiment would be worth repeating on a number of other rhizomatous species. The actual consistency of the soil may be important in influencing the rate of lateral spread. This is shown by the difference in annual increments in the length of rhizomes grown in stiff clay and in open soils. Aeration probably plays a part as well as mechanical resistance to penetration by the growing rhizomes.

Whilst in most rhizomes there is no division of labour between the functions of storage and lateral spread, specialization may be seen in wood-sorrel (*Oxalis acetosella*), where the slender creeping

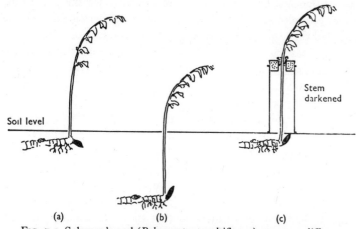

FIG. 7.5. Solomon's seal (*Polygonatum multiflorum*) grown at different soil depths. The direction of further growth of the rhizome is influenced by the length of stem which is darkened. (After Raunkiaer, 1934)

rhizome achieves the dispersal and the food is stored in the fleshy bases of the petioles; in the 'droppers' formed as lateral branches in garden montbretia (*Tritonia crocosmæflora*); and in the thin rhizomes of arrowhead (*Sagittaria sagittifolia*) or potato, bearing tubers at the end where the food storage is localized.

The general vigour with which rhizomatous plants spread, and the dense colonies which they so often form, to the exclusion of other species, may be taken as an indication of the success of their adaptations as a means of agression. Though lateral spread is gradual, the new shoots, with their abundant food reserves to draw upon, can compete more effectively than seedlings. Species multiplying vegetatively quite commonly set few viable seeds (for example, dog's mercury), but as we have seen with yellow archangel, the balance between the two modes of reproduction may be a delicate one, liable to be tipped this way or that by different environmental conditions. The comparison between the relative importance of vegetative reproduction and seeding in the spreading of a species makes a worth-while field study, for it gives real insight into the nature of aggression. One of the difficulties is to establish the extent of individual clumps of rhizome origin in a crowded community. Actual excavation of the rhizomes and tracing their course is bound to be a tedious task, and may be

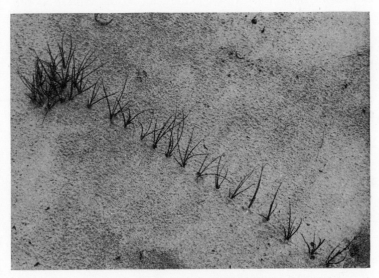

FIG. 7.6. Colonization of sand by sand sedge
(Courtesy of *John Markham*)

very difficult. With some species, like sand sedge (*Carex arenaria*), the growth of the rhizomes is in such regular straight lines that one can often get a fair idea of the extent of individual rhizome-systems by careful mapping (Fig. 7.6). Again, with diœcious species such as dog's mercury, mapping the distribution of stems bearing male and female flowers gives some idea of the extent of vegetative spread, for obviously each male or female patch must have arisen from at least one separate seed. In the case of bracken, especially when growing in dry habitats, the evidence suggests that reproduction from sporelings is rare, so that quite large areas probably represent the vegetative spread of a single plant.

It is clear from any study of plant communities that vegetative spread constitutes a powerful means of aggression. Dispersal by this means must be gradual, but the density of the 'attack' helps to make it effective. In this last point, however, lies one of its drawbacks, for the density of the growth may eventually weaken the aggressor itself through overcrowding and excessive mutual competition. This is seen in the weaker growth of stems in the middle of old clumps of Michaelmas daisies. There are disadvantages also on genetical grounds, for all the vegetative 'progeny' of any plant are virtually parts of a single individual, separated in space and time: they constitute what is called a **clone**. In this there can be no inherent variation allowing adaptation to slightly differing ecological conditions. Thus, however efficient a species may be, if it abandons seed reproduction entirely for vegetative methods it is following a blind alley in evolution — one leading to its own ultimate extinction.

While true dispersal is not normally possible in the vegetative stage, mention should be made here of a few exceptional cases in which it can occur. The most obvious is in water plants, where fragments may break off and be carried away by the current or by water fowl. We have good evidence of the possibilities here in the spread of Canadian pondweed (*Elodea canadensis*). Introduced into the midlands of England in about 1846, it had spread to waterways all over the country within twenty or thirty years, becoming a serious nuisance by its choking of the canals. As the species is diœcious, and for many years only female plants were present in Britain (the male is still very rare), the spread must have been entirely by vegetative means. On land the detachable buds or **bulbils** of such species are lesser celandine (*Ranunculus ficaria var. ficaria*) and some garden lilies can achieve a measure of true dispersal, though this must be quite unimportant ecologically. In the gemmae (buds) of some liverworts (for example, *Lunularia*) we see a similar phenomenon.

8

Sexual Reproduction and Dispersal: (1) Pollination and Variation

In the formation of large numbers of new individuals, and their dispersal as seeds, highly resistant to rigours of temperature and drought, we see one of the most powerful aggressive factors upon which the success of a plant species depends. But there is another aspect of sexual reproduction, less obvious, but no less important for survival — it provides the small variations between the new individuals formed, from which the pressure of natural selection eliminates those less suited to compete in the conditions of the particular habitat. Thus, provided favourable variations arise, the way is open for a gradual trend towards progressive adaptation of the individuals surviving, making them more efficient competitors, both structurally and physiologically, in the conditions under which they are growing — better able to tolerate the difficulties of the habitat and perhaps to oust their rivals. Pollination, as an essential forerunner of seed production and the means by which favourable adaptations can spread through a population, has good claim to be included in this section on aggression.

Suppose that we grow a number of strawberry plants in the garden, all offsets from the runners of a single 'parent' plant, and consequently all with exactly the same genetic constitution (in fact, a clone). Even when grown under good, uniform conditions, the resulting plants will be found to differ in the number of leaves and flowers they produce, and in many other less obvious ways, if they are examined sufficiently closely. With some plant species, one difference which might confuse us is that between the size and shape of *successive* leaves on the same plant (Fig. 8.1) not noticeable in strawberries but very marked in groundsels (*Senecio vulgaris*), harebell (*Campanula rotundifolia*), ivy (*Hedera helix*) and sugar beet. Such progressive changes in leaf shape, if they follow a regular pattern, reflect the stage of maturity which a plant has reached (its 'physiological age'), and it has been suggested that they might enable one to rogue out the varieties of sugar beet likely to 'bolt' early. But the lesson to be learnt from them here is that only *corresponding* leaves should be compared when looking for differences between plant specimens.

Now the differences noticed in our strawberry plants may be partly due to variations in the size and amount of reserve food in

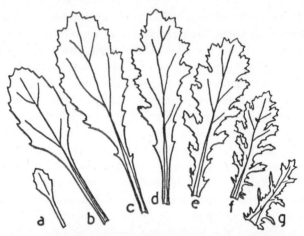

FIG. 8.1. Leaves of *Senecio vulgaris* (groundsel). *a–g*, leaves from successive nodes starting at the base. × ⅔
From W. O. James: Elements of Plant Biology: Allen and Unwin

the individual offsets at the start, and partly to innumerable minor differences in the local environment in which they are growing (such as depth of planting, presence of stones or worm-burrows affecting the texture of the soil in contact with the roots, etc.). But if we were to grow some individuals from the same clone in quite different soils, and some in deep shade we should obviously find much greater differences, and some of these differences we might regard as adaptive. Probably the plants in poor soil, though smaller, would flower earlier than those in richly manured soil, and those in the shade might have bigger leaves (possibly with a larger surface of leaf per unit dry weight, as in bluebells).

With some species we can obtain more uniform planting material in the form of **seeds** with identical genetic make-up, for there are a number of plants in which the vital stages of nuclear fusion and meiosis are short-circuited in flowering and seed production. The term **apomixis** is used to describe reproduction of this kind, where there is no true sexual process (strictly speaking it embraces vegetative propagation as well). Apomictic production of seeds occurs regularly in many of the hawkweeds (*Hieracium* spp.), the common dandelion (*Taraxacum officinale*), hawthorns (*Crataegus* spp.) and very commonly in brambles (*Rubus* spp.). We can see that in such cases the seeds, like the new individuals in vegetative propagation, are produced without any reshuffling of the genes responsible for heredity, and will all be identical in their inherited potentialities or **genotype**. Also they will show much less variation in size than units of vegetative propagation, such as the offsets of strawberries or the tubers of potatoes. Nevertheless seeds from apomictic brambles grown under differing conditions will produce plants differing widely in the size and texture of the leaves, their hairiness and the nature of the prickles on the stem (Fig. 8.2). These differences, resulting directly from the influence of the environment upon the plant's development are called **environmental variations** to distinguish from **genetic variations** brought about by genetic differences. It is, perhaps, not altogether surprising that they should be so marked, for we can see that the general pattern of growth in plants, forming the new organs from the growing point 'as they go along', must lend itself to plasticity in development far more than the highly organized pattern seen in the embryology of animals.

Most species can be found growing naturally in a range of habitats differing considerably in their local environment, and in many cases the appearance of the populations in these different habitats is sufficiently distinctive for the plants to be regarded as separate ecological forms of the same species. The question immediately arises: do these different forms result merely from the effects of the local environment on the development of the plants (environmental variations), or are they adaptive, having a genetic basis (genetic variations)? As well as being of interest to the ecologist, this is a question of the greatest importance in plant classification, and it should be noted here that the impact of ecology on taxonomy is admirably discussed by Heslop-Harrison

FIG. 8.2. The effect of shade on a single genotype of *Rubus incurvatus* shown in seedling leaves and stems. A. Plant grown in full light; B. in medium shade; C. in deep shade. Note the differences in leaf size and stem thickness, and also in hairiness, stem shape in section, and prickle form

From J. Heslop-Harrison: New Concepts in Flowering-Plant Taxonomy, Heinemann

(1953), to whom acknowledgment is due for a number of the ideas and examples in this chapter. The answer to the question is to be sought in transplant experiments which compare the performance of the different forms when grown side-by-side under the same conditions — either the conditions prevailing at the extremes of the tolerance range shown by the species, or in some intermediate environment. If the differences between the ecological forms are due directly to environmental factors influencing the development of the plants, then they should largely disappear when specimens are reared under the same conditions. If, on the other hand, they are the result of different genetic make-up (different genotypes), then one would expect the forms from different habitats to remain distinct, even if somewhat modified, when grown side-by-side in new conditions. The British Ecological Society for many years carried on a series of carefully controlled transplant experiments at Potterne in Wiltshire, and schools could add useful knowledge by similar experiments on a smaller scale. Specimens from different habitats could be grown side-by-side in different soils or with different shading, and all changes carefully recorded. It is essential to preserve good, pressed herbarium specimens (see Appendix) of the plants as at the beginning of the experiment and from each subsequent year, as well as describing any changes observed. So far as possible, detailed records should also be kept of environmental factors, such as shading, pH and water-holding capacity of the soil.

Although a number of 'habitat forms' (environmental variations) are known to occur, the results of many transplant experiments point to genetic differences between the populations of different habitats; the plants remain distinct when grown together under standard conditions. Moreover, the plasticity itself varies in degree and nature between different populations; it is, as would be expected, determined genetically. We can hardly escape the conclusion that these local populations, differing genetically from the general population of the species and occupying particular habitats, must have been selected for their capacity to thrive under the conditions of the local environment. They are in fact adaptive, often obviously so, as in the case of the cushion-like forms common in windswept situations or the creeping form of broom (*Sarothamnus scoparius* ssp. *prostratus*) growing on sea cliffs in Cornwall. Quite commonly, however, the morphological characters by which members of different populations are distinguished may not appear to have any adaptive value. In such cases they may be associated with physiological adaptations, detectable only by careful experimentation, but nevertheless of the utmost importance for survival in specialized habitats. A good example is provided by the sea thrift (*Armeria maritima*), which has forms differing in their tolerance of sea spray. No corresponding morphological differences are associated with this, but it has been shown that the germination of seeds from plants growing

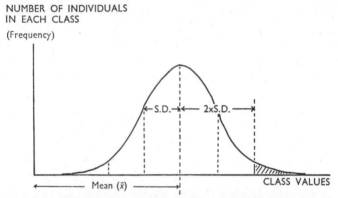

FIG. 8.3. Normal distribution curve. The population concerned is represented by the area beneath the curve. The shaded area makes up about $2\frac{1}{2}\%$ of this, hence only about 5% of the population vary from the mean by more than twice the standard deviation

close to the sea is actually *favoured* by salt solution in concentrations which will greatly reduce the percentage germinations of seeds from plants growing in more sheltered places further up the cliff.

These ecological forms have been termed **ecotypes** by the Swedish botanist Turesson, who was among the first to carry out extensive studies from transplant experiments. We should now consider how they arise, but as a preliminary we must make a slight digression into the realm of statistical analysis of variation, so that we can see how to distinguish between groups whose variants overlap.

If we make a careful examination of any particular character in a plant population growing under fairly uniform conditions, we will expect to find (plasticity apart) some degree of inherent variation in it. We might count the carpels in wood anemones or the petals in lesser celandines, or measure the length of the flower stalk in thrift plants or the spur in orchid flowers. Then, if we arrange our results to show the number of specimens observed in each frequency class (that is, a *particular* number of carpels or petals, or a *particular* range of length, chosen arbitrarily), we should expect the majority of the specimens examined to fall into classes of average value, while relatively few would occur in the low or high classes at either end of the scale. In other words, we should expect our results to conform more or less closely with a **normal distribution curve** (Fig. 8.3). Table 8, giving the results of counts of petal numbers in a sample of 148 lesser celandine flowers (*Ranunculus ficaria*) provides a simple illustration.

TABLE 8. *Petal Number in a Sample of 148 Lesser Celandine Flowers*

No. of Petals (frequency classes)	6	7	8	9	10	11	12	13	14
Frequency observed (that is number of flowers observed in each class)	2	9	80	28	18	7	2	1	1

Now a normal distribution curve is based mathematically on the laws of random probability, and it is characterized completely by two values: the mean ($\bar{x}$), which defines the centre of the distribution, and the standard deviation (*S.D.*), that is, the distance from the mean to the steepest part of the curve on either side (see Fig. 8.3). This gives a measure of the range of variation, in that a definite proportion of the area enclosed by the curve (that is, a definite proportion of the sample population under investigation) falls within the standard deviation (the square of the *S.D.* is called the **variance**). *Roughly speaking, two-thirds of the population fall within one S.D. away from the mean, and 95 per cent within 2 × S.D. from the mean.* It thus provides a very good way of comparing two populations in which the centres of distribution, although distinct, are so near that the ranges of variation overlap.

If we plot our results for the celandine flowers as a histogram (or frequency polygon) (Fig. 8.4), we can see the resemblance to a normal distribution curve, but it is rather approximate. Taking a larger sample would probably do something to improve the likeness, but we can do nothing to smooth the polygon by taking a larger number of smaller frequency classes, since we are dealing only with *whole* numbers — there cannot be fractions of a petal present. When we are handling measurement data (for example, length of flower stalks) the frequency classes are an arbitrary choice, and can be made as fine and numerous as we like, provided the accuracy of measurement and the numbers in the samples justify the choice. It will be seen that something approaching a smooth curve could be produced in this way.

Even with a small sample, and the obvious discrepancies due to the discontinuity of variation when using whole number data, it is often useful to treat such data as though it were a normal distribution. What is done in practice is to calculate from the data the mean and *S.D.* of the normal distribution curve which would give the best fit. The process is quite simple; only the convention of statistical notation may be unfamiliar, and this is best explained by referring to Table 9.

TABLE 9. *Petal Number in a Sample of Lesser Celandine Flowers*

1	2	3	4	5
No. of petals (Frequency class) (x)	Frequency observed (a)	Deviations from Working Mean $(x_m = 9)$ $(x - x_m)$	Frequency $\times$ deviation $a(x - x_m)$	Frequency $\times$ square of deviation $a(x - x_m)^2$
6	2	-3	-6	18
7	9	-2	-18	36
8	80	-1	-80	80
9	28	0	0	0
10	18	$+1$	$+18$	18
11	7	$+2$	$+14$	28
12	2	$+3$	$+6$	18
13	1	$+4$	$+4$	16
14	1	$+5$	$+5$	25
Totals:	$N = 148$		$Sa(x - x_m)$ $= -57$	$Sa(x - x_m)^2$ $= 239$

Mean $(\bar{x})$ = working mean (x_m) + correction $\left(\dfrac{Sa(x - x_m)}{N}\right)$

$$\bar{x} = 9 + \frac{-57}{148} = 8\cdot615$$

Sum of squares of deviations from mean:

$$Sa(x - \bar{x})^2 = Sa(x - x_m)^2 - \frac{(Sa(x - x_m))^2}{N}$$

$$= 239 - \frac{(-57)^2}{148} = 217\cdot05$$

Standard deviation $(S.D.) = \sqrt{\left(\dfrac{S(x - \bar{x})^2}{N - 1}\right)} = \sqrt{\dfrac{217\cdot05}{147}} = 1\cdot215$

Standard error of mean $(S.E.) = \sqrt{\dfrac{(S.D.)^2}{N}}$ or $\dfrac{S.D.}{\sqrt{N}}$

that is, $\sqrt{\dfrac{1\cdot215^2}{148}}$ or $\dfrac{1\cdot215}{\sqrt{148}} = 0\cdot100$

FREQUENCY

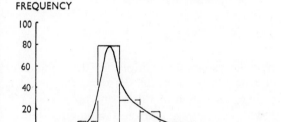

FIG. 8.4. Frequency histogram of petal numbers in 148 flowers of lesser celandine (*Ranunculus ficaria*). (See text)

$\bar{x}$ represents *any* of the petal numbers in the data, that is *any* frequency class in column 1.

a represents *any* of the observed frequencies in column 2, the sum of these being:

N, that is, the total number in the sample (148).

$\bar{x}$ is the true mean of the sample. It would be obtained by multiplying the petal number in each of the frequency classes by the corresponding frequency observed $(a \times x)$, taking the sum of all the products $(S\,ax)$ and dividing by 148 $(S\,ax/N)$. It is often easier to take any convenient number (usually the frequency class in which the highest frequency is observed) as a **working mean** x_m (here we have taken $x_m = 9$) and calculate the correction to be applied. This is done by setting out the deviations from the working mean $(x - x_m)$, as in column 3, and multiplying by the corresponding observed frequency $(a(x - x_m)$, column 4). The correction to be applied is the sum of column 4 divided by the number in the sample:

$$\frac{Sa(x - x_m)}{N} = \frac{-57}{148}$$

The *true* mean, then, $\bar{x} = 9 - \frac{57}{148} = 8\cdot 615$.

The standard deviation is obtained by working out the expression:

$$\sqrt{\left(\frac{\text{Sum of squares of deviations from the mean}}{\text{Sample number minus 1}}\right)}$$

that is, $$S.D. = \sqrt{\frac{S(x - \bar{x})^2}{N - 1}}$$

Since the frequency classes simply multiply up the number of times any particular deviation occurs in the data, $S(x - \bar{x})^2$ is really $Sa(x - \bar{x})^2$. The calculation of this is enormously simplified by continuing with our *working mean* and finding the sum of the squares of the deviations from *this*. $(Sa(x - x_m)^2 = 239$, column 5). By subtracting the squared sum of the deviations from the working mean divided by the number of observations $\frac{(Sa(x - x_m))^2}{N}$ or $\frac{(-57)^2}{148}$ the sum of the squares of the deviations from the *true* mean is obtained. In our example:

$$Sa(x - \bar{x})^2 \text{ or } S(x - \bar{x})^2 = 239 - \frac{(-57)^2}{148} = 217\cdot 05$$

$$\text{Thus } S.D. = \sqrt{\frac{217\cdot 05}{147}} = 1\cdot 215$$

From this we can say that, in the normal distribution best fitting our sample data, some 95 per cent of the population will lie within 2×1.215 on either side of the mean of the sample. This deduction is based on quite a small sample, but from the square of the standard deviation (known as the variance) we can get some assessment of the extent to which our sample data may deviate from the whole population, by calculating the standard error of the Mean (*S.E.*). This is given by the formula:

$$S.E. = \sqrt{\frac{(S.D.)^2}{N}} \text{ or } \frac{S.D.}{\sqrt{N}} : \text{ here } S.E. = \frac{1.215}{\sqrt{148}} = 0.100. \text{ This means}$$

that there is *a probability of about* 20 : 1 *that the mean of the whole population lies within* 2×0.100 *of the mean of our sample, in fact,* within the range 8.615 ± 0.200.

When comparing two populations, not only the mean values for the particular character under consideration, but also the standard deviations, reflecting the range of variation within each population, may differ. Let us examine some data collected from two populations of lady's smock (*Cardamine pratensis*) growing in the same marshy meadow, the one double-flowered and the other the ordinary single-flowered form. Can these forms be distinguished by vegetative characters alone? Measurements were taken of the length of the second leaf from the base of the flowering stem on a number of plants of both kinds (Table 10).

TABLE 10. *Leaf Length in Single and Double-flowered Forms of Lady's Smock*

Freq. class (cm.)	Class mean x	Double-flowered form				Single-flowered form			
		Freq. observed a	$(x-x_m)$	$a(x-x_m)$	$a(x-x_m)^2$	Freq. observed a	$(x-x_m)$	$a(x-x_m)$	$a(x-x_m)^2$
0–1	0.5	1	−3	−3	9	1	−3	−3	9
1–2	1.5	25	−2	−50	100	12	−2	−24	48
2–3	2.5	39	−1	−39	39	21	−1	−21	21
3–4	3.5	10	0	0	0	24	0	0	0
4–5	4.5	7	+1	+7	7	13	+1	+13	13
5–6	5.5	2	+2	+4	8	12	+2	+24	48
6–7	6.5	0	+3	0	0	2	+3	+6	18
Totals:		$N=84$		−81	163	$N=85$		−5	157

For the double-flowered form:

Mean leaf length of sample $(\bar{x}) = 3 \cdot 5 - \dfrac{81}{84} = 2 \cdot 54$ cm.

$$S(x - \bar{x})^2 = 163 - \frac{81^2}{84} = 84 \cdot 89$$

$$S.D. = \sqrt{\frac{S(x - \bar{x})^2}{N - 1}} = \sqrt{\frac{84 \cdot 89}{83}} = 1 \cdot 011$$

For the single-flowered form:

Mean leaf length of sample $(\bar{x}) = 3 \cdot 5 - \dfrac{5}{85} = 3 \cdot 44$ cm.

$$S(x - \bar{x})^2 = 157 - \frac{5^2}{85} = 156 \cdot 71$$

$$S.D. = \sqrt{\frac{S(x - \bar{x})^2}{N - 1}} = \sqrt{\frac{156 \cdot 71}{84}} = 1 \cdot 366$$

To compare the two populations we find the standard error of the difference, given by the formula:

$$S.E. \text{ of Difference} = \sqrt{\frac{(S.D._A)^2}{N_A} + \frac{(S.D._B)^2}{N_B}}$$

where $(S.D._A)$ and $S.D._B)$ are the standard deviations of the samples from the two populations, and N_A and N_B are the numbers in the samples, that is,

$$S.E. \text{ of Difference} = \sqrt{\frac{1 \cdot 011^2}{84} + \frac{1 \cdot 366^2}{85}} = 0 \cdot 185.$$

Now the observed difference between the mean leaf lengths $(\bar{x})$ in the two populations was: $3 \cdot 44 - 2 \cdot 54 = 0 \cdot 90$ cm. But this is far greater than twice the standard error of the difference, hence we conclude that, although the ranges of variation overlap in the two populations, the probability is very high indeed that single-flowered plants have larger leaves, and the differences observed are not merely due to a caprice of our sample.

We must now return from our digression on statistical technique to the question we left unanswered. The evidence of transplant experiments points as a rule to a *genetical* difference between different ecotypes and the main population of the species: how does this arise?

One cannot study the pollination mechanisms of flowering plants without being astounded by the variety of elaborate devices which have been evolved favouring cross-pollination between different plants of the same species, and at the same time minimizing the chances of self-pollination. Not only are flowers adapted structurally to secure this end, but the pollen is commonly shed before the stigma is exposed or receptive (protandry), or in some cases the stigma matures first (protogyny).

Even if, in spite of these measures, some self-pollination does occur, it may still not be followed by fertilization, for as a rule the pollen tubes from such grains grow much more slowly than do those of pollen brought from other individuals. From the simple experiment of enclosing flowers from the bud stage in fine muslin or 'cellophane' bags, one usually finds that little or no seed has set, confirming that the flowers are largely self-sterile. True, there are some species which are habitually self-pollinating, but even among these self-pollination may be restricted to the later formed flowers, like the cleistogamous flowers of violets, which are self-fertilized and set seed without the buds ever opening. In contrast to the self-sterility shown by most species when the flowers are 'bagged' to prevent cross-pollination, counts of the percentage seed-set under normal conditions usually point to a high degree of efficiency in the pollination mechanism. There is considerable scope for applying this simple experimental technique to a number of ecological problems concerning the reproductive rate of different species, as it is hoped the next few pages will show.

Given an efficient system of cross-pollination within a plant population, there should be free gene exchange, with the different combinations of genes arising in random frequencies. Here we see the connection with a normal distribution in a character like size, when it is controlled by a number of different genes (known as a **polygene system**). To take an imaginary simple case; suppose that the length of flower stalk in a particular species is governed by five genes: *A*, *B*, *C*, *D* and *E*, all showing incomplete dominance. If they are all present in the dominant form, that genotype (*AABBCCDDEE*) produces a stalk of maximum length, while the homozygous recessive (*aabbccddee*) gives the minimum length. Obviously, the chances of these two genotypes arising in a freely out-breeding population will be small (like the chances of ten coins, tossed simultaneously, all coming down 'heads'), so that individuals with extremely long or extremely short flower-stalks will be rare. By contrast, plants with about half of the genes concerned present as dominants will have flower-stalks of medium length, and will form the bulk of the population. A normal distribution in the variation of such characters is, then, one indication of free gene-exchange within the population.

Doubtless the wide diversity of cross-pollination mechanisms we see today owe their survival to the enormous evolutionary value of free gene exchange, which allows favourable mutations to spread rapidly through the whole population, giving material on which natural selection can work. At the same time, free gene exchange within a large population gives a valuable measure of stability, by swamping any diverging groups arising from chance combinations of genes.

But in species where distinct ecotypes from different specialized habitats can be recognized, the variation does not conform with a normal distribution — more likely the curve of the whole population would be flattened or even show subsidiary humps. To what extent does this show that there is not free gene exchange within the population? We may surmise that when a species colonizes a new and specialized habitat, its ability to survive the initial difficulties will probably depend upon its inherent plasticity of development. Once it has gained a foothold the pressure of natural selection may well be intense, putting a premium on quite different characters and so favouring the survival of different genotypes from those in the main population.

The hawkweed (*Hieracium umbellatum*) which was intensively studied by Turesson in Sweden provides a good example. The normal form of this species (which is not apomictic) grows in dry sandy meadows, in keen competition with grasses and many other herbaceous plants, but where it colonizes shifting sand dunes we can readily see that its survival will depend far more upon its ability to tolerate the exacting conditions of the new habitat than its powers of competition with other species. Turesson found that this was reflected in differences such as a more erect habit of growth and rapid shoot regeneration in the dune types. These characteristics persisted in culture, confirming a genetic difference between the two ecotypes. This divergence, first imposed by natural selection, is far more likely to become stabilized genetically and assume a unity of its own if there is some breeding barrier giving a measure of isolation from the main population of the species.

The rapid evolutionary divergence of populations on small oceanic islands is well known. In such cases there is complete geographical isolation, but we must realize that breeding barriers have not necessarily the imposing grandeur of a few hundred miles of ocean or a mountain range. Any factor checking cross-pollination with the main population may be effective in isolating a group of plants, which is then likely to show further divergence, though remaining potentially capable of interbreeding with the main population. In short, the breeding barrier will prevent the 'swamping' of particular gene combinations of the ecotype favouring survival in the specialized habitat. We can see, then, that the 'ecology of pollination', which has received all too little attention in the past, offers a fascinating field of study.

Wind-pollination is less subject to local barriers than insect-pollination. Although the chances of any single pollen grain reaching a receptive stigma may be slender, the enormous output of pollen and the ease with which it is carried by the lightest air current, combine to make the mechanism efficient, especially if the species is gregarious, as is true of most wind-pollinated plants (for example, grasses, oaks, birch or pine). In this gregarious tendency we can see a parallel to the communal spawning of such fish as eels or plaice. The absence of large obstructive bracts or petals, the dangling nature of the flowers or their stamens, the smooth surface of the actual pollen grains and the large stigmatic surface to receive them, may all be regarded as products of adaptive evolution favouring greater efficiency. There are also cases of specialized adaptations, such as the air bladders in pine pollen, giving additional buoyancy.

But we are here more concerned with the fact that the pollen is often carried for great distances. Evidence of 'pollen rains' reported from ships at sea, or from the heart of big cities provides substantial proof for this, and it is a fairly simple matter to collect one's own data by exposing sticky slides on the roof of the school buildings (though identification is likely to prove difficult). We are left with the conclusion that there are few local barriers to wind-transport of pollen. Is there perhaps a correlation between this and a lower frequency of ecotype divergence in wind-pollinated species? Incidentally, the fact that the predominant elements in our vegetation (grasses and woodland trees) consist of wind-pollinated species greatly enhances the value of records from fossil pollen deposits in peat, for we know that they are representative of the vegetation for miles around.

The basic adaptations of insect-pollinated flowers lie in the provision of nectar as an attraction for insect visitors, and in the brightly coloured petals which advertise it. The rough, sculptured surface of the actual pollen grains, enabling them to cling to the hairy insect bodies, is also a vitally important adaptation. These refinements have gone hand-in-hand with the evolution of specialized mouthparts in those insects that have come to depend entirely on flower nectar and pollen for their food supply. Furthermore, the range of colour-vision and sense of smell in some groups, such as bees and moths, is closely bound up with the colours and scents of the flowers which they visit. For example, hive bees are colour-blind to pure red, and there are few 'red' flowers which are not in fact various shades of pink or purple, that is, they would appear blue to a bee. Apparent exceptions, like poppies, may look conspicuous to bees if they reflect ultra-violet light strongly, for this is visible to bees. This could be tested, using an ultra-violet lamp (or a mercury vapour lamp) and a pinhole camera with the right kind of film. It is interesting to note that reds and scarlets predominate in tropical flowers habitually pollinated by humming birds, which are said to be colour-blind to blue.

Many kinds of flower have their nectar easily accessible to any visiting insect, and so depend upon a variety of species for pollination. When colonizing different habitats there is a fair chance that the ecological range of some of the pollinating insects will overlap with that of the main plant population so that free cross-pollination may persist. Nevertheless, limitations may be set by windswept situations too exposed for insect life, when the percentage of seeds setting might fall below that necessary to maintain the population. The proportion of insect-pollinated species is certainly smaller in the flora of areas exposed to high winds. Seasonal bad weather may also influence pollination. The flowering period of wood anemones (*Anemone nemorosa*) frequently coincides with a spell of weather so bleak and cold that few pollinating insects can be on the wing, and the percentage of carpels maturing to form achenes is commonly very low — perhaps for this reason.

In specialized flowers the nectar is generally concealed in tubes formed by the fused petals, or by a spur, so that only insects with a long enough proboscis can reach it. There may also be some mechanism as in toadflax or lupin which requires a powerful insect like a bee to work it (Fig. 8.5). As a result, the range of insect visitors is limited to a few species, and the general structure of the flower often 'caters' for these in other ways. Bee flowers always have some kind of landing stage, while those pollinated by hovering moths (for example, honeysuckle) have stamens and style protruding far outside the corolla-tube. Others, like tobacco plants, which are pollinated by night-flying moths are usually white and produce a strong scent at dusk by which they can be detected at a considerable distance. The biological advantage of this specialization presumably lies in improving the chances of cross-pollination. This depends in part upon a valuable feature in the behaviour pattern of the pollinating insects: constancy in visiting one particular species of flower; and in part upon further adaptation of the flower structure whereby the anthers are so placed that pollen will be deposited on that part of the insect's body which will come into contact with the receptive stigmas of other flowers.

Counts of the percentage of ovules setting seed leave no doubt as to the efficiency of these specialized pollination mechanisms. Yet if compared on the same basis, the non-specialized types often show as high an efficiency, and one is left wondering what are the advantages which have led to the selection and perfecting of the specialized mechanisms. There can be no single answer to this question, but in considering any particular case two things should be borne in mind. First, pollination constitutes only *one* of the critical stages in the life of a plant, and we must not forget that the success of any species depends upon how well it can meet *all* the crises which confront it. So, even if pollination is not highly efficient, a species may still flourish because of good seed dispersal.

F

FIG. 8.5. Honey-bee foraging on snowberry (Courtesy of *Joseph Neumann*)

vigorous growth of seedlings, or vegetative spread. Secondly, it does appear that, in general, flowers with highly specialized pollination mechanisms are enabled to economize in their output of pollen, and at the same time have often a relatively large number of ovules per flower. Each ovule in the ovary, of course, needs a separate pollen grain for its fertilization, and it is noticeable that wind-pollinated species rarely have more than one or two ovules per flower, whereas in flowers specialized for insect-pollination the number of ovules may be very high indeed. In orchids, where the pollen output of each flower is transported *en bloc*, the number of ovules per flower may exceed 10,000, so specialization of flower structure does enable these plants to maintain an enormous output of seeds.

But specialization may also bring drawbacks, and must sometimes result in breeding barriers interfering with free gene exchange within the plant population. The trend of specialization is nearly always towards fitting the plant for a limited range of pollinating insects, sometimes a single species. This must mean inevitably that the distribution of the plant species is restricted by the ecology of the pollinating insect, and in habitats where the insect is scarce or absent the plant will not be able to maintain itself. This possibility is suggested by data collected over two successive years on the pollination and seed-set in two colonies of lesser butterfly orchid (*Platanthera biflora*) (*Blundell's Science Magazine*, No. 7). The flowers of this species have their nectar in spurs $1\frac{1}{2}$ to 2 cm. long (out of reach of bees); they are white and heavily scented, especially in the evening. Pollination is by moths, which carry the pollinia sticking to their heads, like an extra pair of antennæ, from one flower to another; thus it is possible to see whether a mature flower has been pollinated or had its pollinia removed. Of the two colonies studied, one was in a sheltered meadow, backed by woodland, in the Exe Valley, while the other grew in boggy grassland bordering on a piece of exposed heath near the Blackdown Hills. Counts were made of the number of mature flowers with pollinia removed, and the number in which the ovaries had developing seed. Self-pollination was ruled out on the basis of a small test in which all the inflorescences that had been enclosed in muslin bags showed no seed development. The results are shown in Table 11.

TABLE 11. *Pollination and Seed Development in Two Colonies of Lesser Butterfly Orchid (Platanthera biflora)*

	Sheltered valley site	Exposed moorland site	Percentage difference
Number of flowers examined:	226	219	
Number of flowers with pollinia removed:	158 (69·9 per cent)	124 (56·6 per cent)	13·3 per cent
Number of flowers with developing seed:	108 (47·8 per cent)	70 (32·0 per cent)	15·8 per cent

We should again introduce a statistical test to confirm the conclusion we draw from what is, after all, quite a small sample of the population. This time we are dealing with **attribute data** (each individual either has, or has not, had its pollinia removed), not measurement data, and a simple test is to calculate the standard error of the percentages observed, using the formula:

$$S.E. = \sqrt{\frac{p \times (100 - p)}{N}}$$

where p = percentage observed, and N = total number of the sample.

Applying this to the figures for percentage of flowers from which the pollinia were removed, we have:

For valley site:

$$S.E. = \sqrt{\frac{69·9 \times 30·1}{226}}$$
$$= 3·05$$

For moorland site:

$$S.E. = \sqrt{\frac{56·6 \times 43·4}{219}}$$
$$= 3·35$$

The chances that the corresponding percentages for the *whole* population differ from those of the samples by more than twice the standard error are less than 1 in 20, and they become *very* remote when the margin is increased towards $3 \times S.E.$ We may say, then, that there is a high degree of probability that the percentage of *all* the flowers in the valley site which had their pollinia removed lies between $(69·9 \pm 2 \times 3·05)$ per cent, that is, between 63·8 and 76·0 per cent. The range for the moorland site works out at between 49·9 and 63·3 per cent. In these particular figures there is no overlap to leave us in any doubt about the reality of the difference between the two populations, but where this does occur, a slightly more sensitive comparison is given by calculating the standard error of the difference:

$$S.E. \text{ of difference} = \sqrt{S.E._A{}^2 + S.E._B{}^2}$$

where $S.E._A$ and $S.E._B$ are the standard errors of the two percentages being compared.

For the figures for the removal of pollinia the *observed difference* is (69·9 − 56·6) per cent = 13·3 per cent.

$$S.E._A = 3·05 \qquad S.E._B = 3·35$$
$$S.E. \text{ of difference} = \sqrt{3·05^2 + 3·35^2} = 4·53$$

The fact that the observed difference (13·3 per cent) considerably exceeds $2 \times S.E.$ of difference (9·06) is convincing evidence that there is a real difference in the proportions of flowers in the *whole populations* at the two sites which were successfully visited by moths — presumably because there were fewer moths on the moor. Although the plants on the moorland site were growing vigorously, and appeared to be competing successfully with the moorland vegetation, their rate of seed production per flower was cut to about ⅔ of that prevailing in the more sheltered valley habitat, and there can be no doubt that a lower rate of pollination was a major factor concerned. Clearly, any further fall in the pollination rate might lead to the butterfly orchids on the moorland site dying out*— a limitation of distribution due to insufficient pollinating insects.

Where the plant is dependent for pollination on very few insect species the numbers of these pollinators may be inadequate for other reasons than habitat-restriction. It is quite likely that very extensive societies of plants may present a pollination task beyond the capacity of the local insect population. The wonderful spectacle of bluebells carpeting an oakwood as far as the eye can see, leaves, as a biological afterthought, the impression that there could not possibly be enough bees to go round, however industrious. It would be interesting to compare the percentage seed-set

* In 1956 and 1957, five years after these observations, scarcely any specimen could be found on the moorland site.

in such a stand with that of a smaller patch more obviously within the capacity of the local bee population. As such a close stand of plants must almost certainly have been built up by vegetative reproduction, any falling off in the rate of pollination, if it does occur, is unlikely to have much effect on the community. Fluctuations in insect populations, however, such as are well known in butterflies and moths could have profound effects on plant survival. It is said that the greater bindweed (*Calystegia sepium*) seldom sets fertile seed in Britain because the convolvulus hawk moth, upon which it depends for its pollination, has become so rare. This may not be a valid example as the flowers are freely visited by humble-bees; in any case the plant maintains itself very adequately by vegetative reproduction.

Finally, a rather trivial, but interesting risk from specialization arises when the pollinating insects (for example, bees) discover that it is easier to bite a hole in the outside of the nectar pouch than to work the pollination mechanism of the flower in the usual way. On more than one occasion when this has happened to the author's runner beans there was not a single unpunctured flower to be found. Happily the bean crop was unaffected, from which one deduces that runner beans must be self-fertile.

Bearing in mind the localized distribution and movement of individual colonies in some species of butterflies one can hardly escape the conclusion that ecological barriers *to the pollinating insects* must sometimes bring a measure of breeding isolation between neighbouring populations of plants. A stretch of wind-swept grass, a narrow strip of woodland or even a tall hedge may restrict the movement of some insects and possibly set up a 'pollination barrier'. Even if the isolation is not complete (there could well be some mixing of the plant populations through seed dispersal) it might be sufficient to allow a degree of genetical divergence, especially if the habitats differed on either side of the barrier. Another way in which a pollination barrier might arise is by the dispersal of the plant population to a new habitat avoided by the original pollinating insects, but where some other suitable insect species are available as pollen vectors.

These ideas about pollination barriers must remain as speculations until evidence is available confirming their reality. One species worth investigating is cuckoo-pint or lord-and-ladies, in

which there is a nice adjustment between the ecological requirements of the plant and the insect which carries its pollen. Growing
usually in deep shade, there would be little chance of visits from
bees or butterflies, even if the flowers were attractive to them, but
instead, they are adapted to attract small, shade-loving psychodid
flies. The attraction is largely due to the foetid smell of the spadix
(as may be shown by enclosing some freshly cut pieces in a box
trap, with a glass funnel to serve as an entrance); the dull red
colour may also play a part. The elaborate mechanism by which
the visiting flies are detained until the staminate flowers have shed
their pollen is well known, and counts of the proportion of ovaries
setting seed point to a high degree of efficiency. In one sample of
50 inflorescences 89 per cent of the ovaries developed: the flowers
are self-sterile. Now these little psychodid flies are characteristic
of moist, sheltered habitats, and, as their range of flight is probably
quite limited, one would imagine that relatively narrow stretches
of open, sunny ground would constitute an ecological barrier for
them. Of course, they might be wind-blown for considerable
distances, but, like aphids, wind may discourage them from flying
at all. If open ground is for them an ecological barrier, as seems
possible, the lords-and-ladies plants should be affected in two
ways: (1) isolated plants left on recently cleared ground, and some
distance from cover would probably not be pollinated, or, at least,
show only a low rate of fruit formation; (2) colonies of the plant
growing in any isolated copse are likely to be in breeding isolation,
and may well show genetical divergence if examined sufficiently
closely. The spotting of the leaves, which is variable, might serve
as a distinguishing character (see Prime 1960).

Of course, breeding isolation is well-recognized where distances
are greater, even though there may be no actual ecological barriers
to the pollinating insects. Where a plant species is distributed
continuously over a fairly wide range, free gene exchange obviously
can take place only within comparatively small sections of the
range, and widely separated sections are in breeding isolation.
Different gene-complexes with the best local survival value, which
are selected by the varying environments along the range, can
therefore be stabilized. Thus we often find a continuous gradient
of ecotypes matching the gradient in ecological conditions.
Populations showing such *continuous* variation (having a genetic

basis) are termed **clines**, and where the variation can be correlated with ecological conditions — **ecoclines**. This is in contrast to the discontinuous variation arising from genetically isolated pockets of plants in the population. Ecoclines are doubtless of common occurrence on mountain slopes where species show any considerable range of altitude, and examples have been studied intensively in the United States. In Britain, a detailed study of the sea plantain (*Plantago maritima*), carried out by Gregor at the Scottish Plant Breeding Station, revealed a clear ecocline (Heslop Harrison, 1953). Specimens taken throughout its range, from the waterlogged mud beyond the saltmarshes to the drained coastal mud further inland, and reared under standard conditions showed a steady gradation in their habit of growth.

The final kind of pollination barrier may arise through isolation in time. Here again, factors which could easily pass unnoticed may constitute an effective barrier; for example difference in the time of day that the pollen is released or the stigmas are receptive. Two closely allied species of bent grass, *Agrostis tenuis* and *A. canina* are prevented from interbreeding in nature by isolation of this kind (Heslop-Harrison, 1953). More obvious cases of isolation in time are the results of flowering at different seasons of the year, which may be initiated by differing responses to length of day. Plants responding to short days may flower either in the spring or the autumn, and one is tempted to think of the distinction between species like spring and autumn squill (*Scilla verna* and *S. autumnalis*) as originating from chance isolation in time in this way.

It will be seen that here lies a field of ecology that has been little studied as yet, where valuable results can be obtained with no more equipment than a pencil, notebook, keen observation and a good fund of patience — a great opportunity for schools. Percival (1965) is a mine of information on floral biology and contains many valuable suggestions for practical work.

We have been concerned in some detail with *pollination* barriers and their effects on variation, but we must not overlook the fact that many varieties owe their origin to chromosome 'accidents' which themselves commonly impose *breeding* isolation. Genetic isolation, though undoubtedly the basis of most specific differences, comes outside the province of ecology, but one particular kind — **polyploidy** — must be mentioned because of its effects on the per formance of plants. It is estimated that about half the species growing in the north temperate regions of the world are polyploids. Polyploidy arises by a doubling of the two sets of chromosomes present in the normal individual, giving rise to a new organism with four sets, and often endowed with greater vigour and wider tolerance than the diploid from which it came. The greater hardiness of polyploid grasses is reflected in their prevalence in certain arctic regions and in deserts where they have been studied. Another kind of polyploidy (**allopolyploidy** as opposed to **autopolyploidy** just described) arises from a doubling of the chromosome complement of hybrid crosses which would otherwise be infertile. This is well illustrated by the rice-grass mentioned in Chapter 1. Of the two species from which it arose, *Spartina maritima* has a diploid chromosome number of 56, and *S. alterniflora* 70. In the formation of the hybrid *S. townsendii*, which has proved such a vigorous new species, there was evidently no reduction division, and a polyploid resulted with a normal chromosome complement of 126.*

* Note on reprinting, 1965: In the light of recent work by Marchant at Kew the situation is not so simple, and it seems the original chromosome counts (made using early techniques) were unreliable. The parent species are now given as: *Spartina maritima* (=*stricta*) $2n = 60$, and *S. alterniflora* $2n = 62$. *S. townsendii* ($2n = 62$) exists in a male-sterile form reproducing vegetatively, and a fertile form ($2n = 124$). Other forms of *S. townsendii* with $2n = 120$, 122, 90 (60 + 30) and 76 (45 + 31) have also been found in Southampton Water.

9

Sexual Reproduction and Dispersal: (2) Fruits and Seeds

It is an obvious necessity that reproduction in either plants or animals must be followed by some means of dispersal. For animals this presents no problem, as even the few sedentary forms have a motile dispersal stage in their life-history. Plants must normally depend upon external agencies like wind, water and animals, and their adaptations for dispersal are directed towards exploiting these to the best advantage. But even with these refinements, there still remains a large element of chance in the dispersal, a factor which must not be forgotten in studying plant communities.

The earlier land plants (like present-day cryptogams) were dispersed as spores, consisting usually of a single cell, so that opportunities for adaptation were limited. Devices for ejecting the spores, like those of modern fern sporangia were doubtless in existence. These tiny spores can be carried for great distances by the wind, so that their actual dispersal is effective enough, but of no avail unless followed by successful germination. Here lies the weakness, for, with only a minute food reserve available in the spore the gametophyte stands a poor chance of establishing itself in face of competition, and may be still further hampered by intolerance of dry conditions.

With the evolution of heterospory and the retention of the megaspore in the ovule, the original means of dispersal was lost to Gymnosperms and Angiosperms, but was replaced by a new structure for broadcasting the species — the seed. The problem of dispersing the microspores also changed, for, as pollen, they must reach a particular destination: the stigma of another flower. Seeds offered great new possibilities, for, while retaining a high degree of resistance to desiccation and extremes of temperature, they could carry much larger food reserves than spores, and being multicellular lent themselves to elaborate adaptations giving better dispersal. The greater complexity of the fruit, as compared with the sporangium, also offered possibilities of more varied adaptation for seed dispersal. We see then in the fruit and seed an example of what Julian Huxley has termed a 'new piece of biological machinery', with the adaptive radiation that has followed on it.

Before looking at the general problems of seed formation and dispersal let us be clear about the functions of seeds. Their primary function of permitting dispersal serves to reduce competition with the parent plant and between fellow-seedlings; they also enable the species to multiply, to colonize new areas and extend in range — often a most important means of aggression. At the same time dormancy provides a resistant stage which can save the species from extinction by adverse conditions, such as cold or drought, and with many species showing 'intermittent germination' it allows dispersal in time. This acts as an insurance against a disastrous season wiping out the whole population, for those seeds which did not germinate in their first year are still viable in the ground to carry on. Finally, we must not overlook the function of germination and the successful establishment of the seedling. This is without doubt the most critical stage in the life-cycle of any plant, for we know that seedling mortality is very high indeed — though, oddly enough, there is very little detailed information available about its exact causes.

For any individual plant at the fruiting stage there is a limited total amount of food material available for seed production; but this may be used in very different ways. If it is concentrated in relatively few large seeds, these will have a much greater chance of successful germination in competition, for the seedlings can grow to a considerable size on their food reserve before they need be self-supporting. But the seeds are few in number and difficult to disperse. At the other end of the scale the emphasis is on numbers. A large number of very small seeds may be produced, making dispersal easier and more widespread, but greatly increasing the risks at germination. This also involves specialized pollination, as large numbers of ovules must be fertilized. That we find in nature every possible stage between these two extremes need not surprise us, for different ecological niches favour different solutions of this problem.

Even though in annual species food material for the seeds forms a high percentage of the total dry-weight of the plant, there must obviously be much less available than in the larger perennial species, particularly trees. The annual species are typical of open habitats, where their seedlings, year by year, have a far better chance of establishing themselves than in a closed community. Characteristically they have small seeds, which by good dispersal *avoid* heavy competition by 'finding' habitats which are free from it. The seed output is fairly high, allowing the species to stand a high mortality among seeds which do not reach suitable open habitats. This is also true of many perennial herbs growing in open ground, while those of more crowded communities tend to form seeds which are not so small. Trees generally produce larger seeds. These have less chance of reaching the ideal habitat, for their dispersal is poorer both by reason of their size and often smaller numbers. Furthermore even with good dispersal, there would be little chance of them escaping competition from their parents where these grow in closed communities as in the case of woodland trees. Their seeds, then, tend rather to *meet* competition as it comes with their own ample food reserves. We have already noticed a relationship between shade tolerance of seedlings and the degree of shade cast by their parents in some species of trees.

Salisbury (1942) has made extensive studies on the reproductive capacity of plants, and the author wishes to acknowledge that this work has been freely drawn upon in this chapter. Marshalling evidence from nearly 300 species, Salisbury has shown a very definite broad relationship between the seed weight of a species and the kind of habitat, or stage in the succession (see Chapter 10) in which it normally grows. This is far more precise than the extreme cases mentioned above would indicate. A steady increase in the average weight of seeds is observable from herbs of open habitats; of short turf; of meadows; and of scrub; to woodlands; shrubs; and trees. Also, in the case of a number of closely allied species from varying habitats he showed that the species from the more closed communities have in each case the larger seeds (Fig. 9.1). Of particular interest is a class of plant, not fitting into the general series, which he called the 'opportunists'. They are colonizers of intermittently open habitats, such as the ground exposed to the light after a fire, or the felling of a tree, or by a

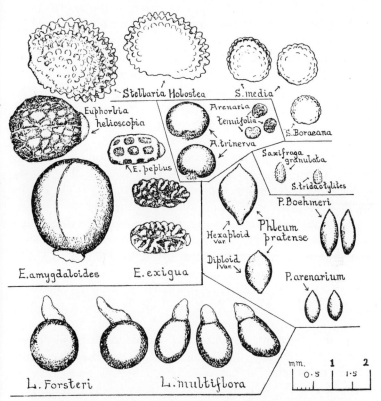

FIG. 9.1. Seeds of species of open and more or less closed communities belonging to the same genus. Note the smaller-sized seeds of the open habitat species. Chickweeds: *Stellaria boraeana*, a species of sand dunes; *S. media*, of cultivated ground and semi-open habitats; *S. holostea*, a woodland plant. Sandworts: *Arenaria tenuifolia*, a species of open habitats; *A. trinerva*, a woodland species. *Saxifraga tridactylites*, a dune species, and *S. granulata*, a species of meadows. Spurges: *Euphorbia exigua*, *E. peplus* and *E. helioscopia*, species of cultivated ground, and *E. amygdaloides*, a woodland species. Cat's-tail grasses: *Phleum arenarium*, a sand dune species; *P. boehmeri*, a species of short turf; *P. pratensis* a meadow species. Woodrushes: *Luzula multiflora*, a species of light patches and coppiced woodlands, and *L. forsteri*, a species of partial shade

From Salisbury: Reproductive Capacity of Plants, G. Bell and Sons, Ltd.

pond drying up and they are able to do this by a phenomenally large output of minute seeds giving them wide and efficient dispersal. Their seeds *require* light for germination. Good examples are: foxglove (*Digitalis purpurea*), mullein (*Verbascum thapsus*) and rosebay willow-herb (*Chamænerion (Epilobium) angustifolium*).

Growth of this last is particularly favoured by the ash left in the soil after a fire, and its regular appearance on burnt ground has earned it the name of fireweed.

Turning to the *numbers* of seeds produced, we are more concerned with the effects of soil conditions and competition on the output of the individual plant. The difference in performance between a luxuriant and an impoverished specimen can be quite astounding. Salisbury (1942) gives figures for counts made on specimens of hairy bittercress (*Cardamine hirsuta*) collected from sand dunes, and from clay soil on the top of a wall bordering the dunes. The average output of seeds from samples of about 100 plants from each soil was approximately 98 seeds per plant in the dunes and 640 seeds per plant growing in the clay. One particularly luxuriant specimen growing as a weed in well-manured garden soil nearby was estimated to have a seed production of nearly 52,000! Here the reduced seed production of the dune specimens resulted from poor soil conditions, but heavy competition has been shown to have a similar effect. Both the number of fruits and the number of seeds per fruit (in capsular fruits) are liable to suffer reduction. Even on extremely impoverished specimens, however, the size and vigour of *individual* seeds appears to be unaffected. We can see, then, that the capacity of a species to maintain or increase its numbers is profoundly influenced by soil conditions and competition for light.

One lesson to be learnt from this is the extreme caution which is needed in estimating the average reproductive capacity of a species. One is tempted to make counts of the seed output from a number of *averaged-sized* individuals (judged by 'eye'), but as Salisbury points out, this can be very misleading, for though conspicuous, these often form but a very small proportion of the population, which as a whole is made up largely of rather weakly specimens. There is, then, no safe short cut: one must make counts of fruits on a considerable number of specimens of all sizes, taken at random *in their natural habitat*, arrange the results in frequency classes based on the number of fruits per plant, and then work out the mean (as we did for the petal number in lesser celandine, p. 143). The mean number of seeds per fruit is determined in the same way, and from these two the average output of seeds per plant in the population can be estimated. But we have not really finished yet, for we have no right to assume that all the seeds are viable, and indeed tests show that for many species a fair proportion are useless. So tests must be carried out with samples to estimate the percentage germination. It is by no means easy always to obtain a reliable figure for this, as some of the seeds which fail to germinate in the tests might do so in the following year, and again, the germination rate may be affected if external factors such as light and temperature are not optimum. We must usually be content, then, to base our estimates of the reproductive capacity of a species on seed counts alone, but they must be carried out on plants growing in their natural habitat.

Salisbury's conclusions, based on a very large number of observations, are that there appears to be little relationship in general between the average reproductive capacity of a species and the risks of mortality to which it is normally exposed. For example, annuals and other plants of open habitats, having small seeds with a consequent high mortality risk, do not as a class have a high output of seeds. He did however, find a correlation between seed output and the frequency or rarity of a species. Comparing closely related species, there is a marked tendency to higher seed output by the ones having the widest tolerance and distribution (hence the commonest), while those with a more limited ecological range produce fewer seeds. The squills and bluebells provide a nice example. Bluebells (*Endymion nonscripta*, formerly included in *Scilla*), which occur in practically every county of the British Isles and can grow successfully on a wide range of soils, show a far higher reproductive capacity than spring squill (*Scilla verna*), which has a distribution limited to about one-quarter of the total counties and vice-counties. Finally, autumn squill (*Scilla autumnalis*), with a restricted southern range (it is entirely absent from Scotland and Ireland), and tolerant only of light soils, has much the lowest reproductive capacity of the three. The important inference is that it is not just the higher seed output which has made bluebells more widespread, but that a reproductive rate much in excess of normal mortality risks has survival value only if there are fair chances of establishment, that is, if the species has a fairly wide tolerance of ecological conditions.

Some trees, notably beech, show a wide fluctuation in the output of fruits or seed in successive years, apparently having no correlation with weather conditions. These 'mast years' may have a survival value in ensuring the periodical establishment of seedlings for regeneration, and are reflected in the populations of saplings of even age sometimes seen in semi-natural woodland. A continuous high output of fruits would only encourage a larger animal population to destroy them and the resulting seedlings.

We have seen that larger seeds, while offering better chances of successful germination in competition, carry with them the disadvantage of being more difficult to disperse. The survival value of wide dispersal is brought home to us with tremendous emphasis when we examine the intricate adaptations which have been evolved to secure it for seeds of all sizes. These must be considered in relation to habitat and the ecology of the individual species if we are to understand them properly. For the most part they exploit few external factors: air and water currents, and the movements of animals, but there is a wealth of variety in the ways of doing it which transforms the rather dull, formal classification of fruits into a fascinating study.

The simplest way of gaining extra advantage from air currents is by an increase in the surface/volume ratio of seeds by hairs (as in willows, poplars, the willow-herbs), or by membranous wings or frills (pine, garden *Nemesia*). An essential function of dispersal is the separation of fellow-seeds, but where the carpel contains only one seed (achenes), or the ovary splits into one-seeded portions (schizocarps), the fruit wall is commonly dispersed with the seed and can be adapted to form the wing (elm, ash, sycamore). Structures even further afield may be involved, as the calyx forming the parachute bristles of dandelion and goat's beard (*Tragopogon*), or the bract which remains on the stalk of lime fruits. These devices give to smaller seeds the widespread ubiquitous dispersal required by opportunists and other plants colonizing open habitats. The larger winged seeds and fruits (and the small ones of willows and poplars) are more typical of trees with seedlings intolerant of shade. In dense woodland communities, this kind of dispersal is of no avail, and the species tend to die out.

The physics of these structures is worth more than passing attention (*Blundell's Science Magazine*, No. 9) as giving some insight into the nicety of the adaptations: also it makes a refreshing change to apply the dynamics equations to dandelion and sycamore fruits. The results are interesting. Even in a fall of $4\frac{1}{2}$ metres, dandelion fruits do not reach a steady terminal velocity, but show a small, approximately constant acceleration of about 6 cm. per sec.^{2}, apparently due to a slight 'streamlining' of the bristles as the rate of fall increases. Perhaps these results may seem to have little relevance, as dandelions do not grow on stalks four and a half metres high, but there does follow from them the fact that an upward eddy-current of little more than 1 m.p.h. is sufficient to keep the fruit floating. If we think back a few years to our childhood's game of blowing dandelion 'clocks', we realize that the minimum wind necessary to detach the fruits may well provide eddy-currents sufficient to keep the fruits floating for quite a time. Sycamore fruits (actually mericarps, or parts of a scizocarpic fruit) achieve the parachute effect in a more complex way. The wing has a thickened leading edge, tapering towards the rear, which brings its centre of gravity well forward. When the fruit is allowed to fall, the position of equilibrium is such that the thickened edge of the wing is lower than the tapered edge. As it falls there is a resultant force acting on the wing, and since the mass of the wing is almost negligible compared with that of the seed, rotation takes place about the seed. Since the dip on the wing is only about 10–20°, the upthrust is virtually unaltered, and the rotation serves merely as a stabilizing device. In this case the rate of fall of the fruit should be roughly the same as that of a weight suspended beneath a piece of paper having an equal area to that of the wing, the total weight of the model being the same as the weight of the fruit. Tests with models showed this to be approximately true, though the actual sycamore fruits had a rather slower rate of fall. Estimates of the mean rate show that theoretically a 20 m.p.h. wind would disperse the fruits for a distance of about five times the height from which they have fallen, and give a spread of about 20 per cent between individual fruits due to differences in seed-weight and wing-dimensions. The dispersal mechanism of ash fruits is much inferior to that of sycamore as the fruits fall at approximately double the rate.

More subtle use is made of the wind by censer mechanisms which form one of the commonest devices for seed dispersal in herbaceous plants. Their operation depends basically upon movements of a springy, upright stalk supporting the ripe fruit capsule.

As this swings to and fro in the wind it produces a centrifugal force on the seeds, which, if large enough will throw them out of the opening at the *top* of the capsule, at the same time giving a forward component to their movement so that they are scattered some distance from the plant. A neat control is achieved, because no release of seeds can take place unless the wind is strong enough to cause considerable bending of the stalk. Theoretical considerations show a *minimum* dispersal of $\sqrt{2} \times$ height of the fruit-stalk, which was in broad agreement with experiment, although some seeds fell short of this range. One requirement of this mechanism is that the seed capsules should be upright, a position that is in some cases only reached as a result of marked post-floral movement of the flower-stalks (for example, in bluebells). Many capsules (for example, red campion (*Melandrium rubrum*), snapdragons) close by hygroscopic movements in damp weather so that the seeds are not released when they would get wet and effective scattering would be hampered by surface tension forces. Censer mechanisms do not give the same widespread dispersal as that achieved by airborne 'floating' seeds, but tend to produce groups of plants which can advance their frontiers quite rapidly under favourable conditions.

Adaptations to dispersal by water are usually inconspicuous. All that is needed structurally is some means of trapping air so that the seeds will float for long enough for water currents to disperse them adequately. The tissues of most water plants are in any case buoyant owing to the large intercellular spaces, so that this need not present any great problem, but special corky structures are present on the seeds and fruits of some marsh and water plants, for example, gipsy-wort (*Lycopus europæus*), and in many sedges (*Carex* spp.) the bract enclosing the fruit forms an air pocket. Physiologically, these seeds must be able to retain their viability after long immersion in water.

There is no need here to go into any details of the varied structures which are involved in succulent fruits to exploit dispersal by animals. As in the case of pollination, both food-reward and advertisement by bright colours are used to attract the animals. The widely scattered distribution of rowans and hawthorns in the rough grassland bordering heather moors is eloquent proof of the efficiency of the dispersal of their fruits by birds (Figs. 1.1 and

9.2). Obviously in some cases dispersal over great distances must be possible. Although birds are the usual vectors, it is said that certain seeds and fruits (for example those of violets and bugle) are carried around by ants for the oily outgrowths (elaiaosomes) which they bear.

A variety of structures have also become involved in adaptations for bur fruits — hooked bristles on the fruit-wall (enchanter's nightshade; goose-grass); hooked styles (wood avens); hooked

FIG. 9.2. Chalk scrub, Chiltern Downs. Stage in succession. Animal dispersal of seeds (Courtesy of the *Forestry Commission*)

involucral bracts (burdock); the rough awns on the bracts enclosing grass fruits. These adaptations can serve no useful purpose save in habitats where there is a suitable population of mammals (or birds), and they are most characteristic of plants of scrub vegetation and the woodland margin. The dispersal which they achieve will tend to follow animal paths, human and otherwise. Careful mapping of the badger tracks in a wood and the distribution of burdock plants should prove an illuminating study. In burdock the dispersal is more effective than in most other bur fruits, for the whole involucre becomes attached to the animal's fur, and the numerous achenes which it contains are shaken out separately along the track. It would be interesting to know just what difference the scarcity of rabbits is making to the dispersal of species like agrimony, goose-grass and enchanter's nightshade.

An aspect needing more investigation is the dispersal of seeds which become sticky when wet due to the presence of mucilage in their coats. This property is so inconspicuous that it can easily pass unnoticed, but there is no doubt that it may play an important part in the spread of seeds which possess it, for example, those of plantains and flax. Plantain seeds, when wet, are very easily carried around on agricultural or garden implements and on one's boots. Bird transport may well play an important part here, and at times seeds may be carried for very great distances. This brings to mind Heslop-Harrison's convincing suggestion that blue-eyed grass (*Sisyrinchium angustifolium*) has been introduced into western Ireland from the United States by Greenland white-fronted geese, whose migration routes, both from the United States and from Europe converge on west Greenland.

Finally there are a number of specialized mechanisms by which plants spread their seeds themselves, without outside help. Some of these are by way of being oddities, and one must guard against getting exaggerated ideas of their importance because of their 'news value'. The basic principle underlying the mechanism of most of them lies in the different degree of shrinkage in length and breadth of fibres on drying. The fruit walls of legumes (vetches, lupins, etc.) are made up of two layers with the fibres running diagonally, but on a different diagonal in each layer so that they are set approximately at right angles.

Differential shrinkage on drying gives a spiral twist (which is quickly reversed if the twisted pods are soaked in warm water) and can be imitated by gumming together two strips of paper cut on different diagonals and drying them over a lamp. The 'explosive' mechanism of such species as gorse, sweet peas, etc., depends on the two valves of the pod tending to twist in opposite directions as they dry, but being prevented from doing so by the tissues of the seams holding the two together. As the twisting force builds up with progressive drying out of the fibres, there comes a time when the resistance of the seams suddenly gives completely and the seeds are flung out as the valves take up their twisted position. By clamping one valve of a sweet pea fruit at the upper end and timing the oscillations when a 200 gm. weight was firmly attached to the lower end it was possible to calculate the restoring couple per unit angular displacement, and from this the kinetic energy given to the seeds on dehiscence.

A theoretical estimate of the range worked out from these considerations came to about 270 cm. for a fruit dehiscing 50 cm. above ground-level. In actual tests with several fruits over sifted earth most of the seeds did fall between 200 cm. and 300 cm., though some were scattered more widely, doubtless due to more favourable placing of the seeds at the end of the pod. The maximum range in the tests was 3½ metres, showing that the mechanism is a very effective means of dispersing quite large seeds. The dehiscence of some siliquas of Cruciferous species (for example, hairy bittercress (*Cardamine hirsuta*)) works on the same general principles, though differing in detail. In some of these dispersal may be further helped by the flattened seeds being able to 'glide' a little. In the capsules of Indian Balsam (*Impatiens glandulifera*), an introduced species rapidly gaining ground in the west, a similar explosive mechanism depends upon tissue tensions set up by turgor in the fruit walls.

A rather different principle is seen in the fruits of wood sorrels (*Oxalis* spp.); it is particularly well developed in the introduced garden weed with bronze-coloured foliage, *Oxalis corniculata*. The capsule seams open without violence, but each seed is held in a purse-like elastic case (an arillar structure) which squeezes out the seeds with such vigour that they may be thrown at least three or four feet. The mechanism is illustrated by an ungentlemanly practice in which most of us have probably indulged in our youth: that of projecting cherry stones by squeezing them smartly between the thumb and finger (representing the elastic aril). When ripe, the mechanism is triggered off by the slightest touch or movement.

DORMANCY AND GERMINATION

In considering seeds one often tends to think about them only in terms of their structural adaptations for dispersal, and it is easy to forget about the physiological adaptations which play just as important a part in fitting them for their role. They are more than just tiny plants, conveniently compact and parcelled up with a food ration, ready for transport. In their state of **dormancy** they are physiologically adapted to stand up to the rigorous conditions they may meet on their journey, when they will usually have no access to a water supply, and temperatures may be extreme. Prolonging this state of dormancy is clearly of great biological value, for the resistant seeds are able to come unscathed through the worst catastrophes of drought, heat, winter cold or disturbances of the soil by cultivation, which would wipe out the adult plants. Finally, having survived these crises they must be able to germinate and establish the seedlings in their new homes.

The resistance shown in the dormant state is normally associated with a low water content in the tissues, especially in the actual protoplasm. Seeds, in fact, appear to carry to a stage of greater perfection the physiological adaptations which we have already noted in winter buds and xerophytes. But it is the variations on this general theme which interest us here. There are seeds which can remain dormant only for short periods: those of poplars and some species of willow will die if they have not met with suitable conditions for germination within one or two weeks. With very short-lived seeds the resistant stage does no more than tide the seed over the rigours of actual dispersal, and is of little or no value in enabling the species to survive subsequent periods of winter cold. The examples mentioned above are well able to meet these emergencies with their hardy winter buds, and there can scarcely be said to be any risk of their wholesale extinction by a disastrous season. The same cannot be said of dune ephemerals, and the annual weeds of walls and arable land. Many of them are able to germinate early in the spring and complete their life-cycle before the ground has dried out. Hairy bittercress (*Cardamine hirsuta*), rue-leaved saxifrage (*Saxifraga tridactylites*) and some of the mouse-ear chickweeds (like *Cerastium semidecandrum*) are good examples.

For many kinds of seed there is a period of 'enforced' dormancy before germination can take place, even if conditions are favourable. This may be due to the fact that the embryo is not fully developed at the time the seeds (or fruits) are shed. Such is the case in buttercups (*Ranunculus* spp.) and wood anemone (*Anemone nemorosa*) where development of the embryo, at the expense of the food reserve, proceeds very gradually during the late summer and into the winter, and germination cannot take place until a definite stage has been reached. More often the cause lies in the relative impermeability of the seed-coat to water (for example, water plantain (*Alisma plantago*)), when germination can be hastened artificially by chipping the seed-coat, a trick often used by gardeners to germinate sweet peas or *Canna*. Presumably the natural processes of decay in the soil achieve the same end in time. Fairly recently it has been shown that the presence in the seed-coat of traces of chemical substances, known as germination inhibitors, is commonly responsible for delaying germination. Under natural conditions these are washed out in time, and germination can proceed when external factors are favourable. At first sight this seems a strange arrangement, but it is one of the underlying mechanisms of a most interesting adaptation, that of **intermittent germination**, which secures *dispersal in time*. By this we mean that the seeds produced by a plant in any one season do not all germinate at the same time, even when they are kept under similar conditions. Not only does this reduce the mutual competition between seedlings, but it has the important survival value of saving populations of delicate species from extinction in a hard year, for there are always some ungerminated seeds left in the ground to carry on. Intermittent germination is probably a common phenomenon, though it varies greatly in degree and is by no means universal. For example, the seeds of the ephemeral species mentioned above, germinate simultaneously as soon as conditions permit. Clearly, intermittent germination is of greatest value for southern species which can only just maintain themselves in our climate. The annual rock rose (*Helianthemum guttatum*) which shows this property is just such a species.

The time lag in intermittent germination may arise from impermeable seed-coats or from germination inhibitors. In some species (for example, corn gromwell (*Lithospermum arvense*)), germination of a batch of seeds may continue irregularly over a long time, but in others, including the rock rose just mentioned, it is discontinuous, occurring in bursts well separated in time. Two distinct kinds of seed may be formed in such cases, like the soft and hard seeds of clovers. The soft seeds soon germinate in favourable conditions, but the hard seeds may remain dormant for years, though if their seed-coats are removed they will germinate at once. This formation of different kinds of seed lends further support to the interpretation of intermittent germination as an adaptation which secures dispersal in time.

One other function of germination inhibitors is to stop premature germination, before the seeds have been dispersed. Without some restraining influence this might well occur in damp weather; even the conditions inside a ripe berry fruit might be favourable. In excessively wet seasons the corn may start sprouting while still in the stook — presumably in such cases the grains have been so thoroughly wetted that the germination inhibitors have already been leached out.

When the period of enforced dormancy is run out, be it long or short, then germination will proceed if conditions are favourable. The successful establishment of the new seedling is one of the most critical stages in the plant's life-cycle, and it is clear that only a very small proportion survive. Basically the conditions required for germination are an adequate supply of moisture, suitable temperature, and enough oxygen for respiration; also the concentration of carbon dioxide must not be excessive, else it will act as an inhibitor. But the variations in the degree of these conditions which may be regarded as favourable for different species may be of considerable ecological importance. Clearly, modest water requirements for germination must characterize dandelion fruits, and be a potent factor in helping them to strike and establish themselves on a gravel path. Low germination temperature in dune ephemerals must be a key attribute enabling them to start their spring growth early. Further, temperature requirements must have a profound effect on the time of germination of seeds buried deeply in the ground, as soils warm up so slowly in the spring. The temperature factor is highly complicated by the effects which periods of low temperature experienced by the seed *during* wet dormancy in the ground, have on the actual germination and subsequent development. For some seeds (*Adoxa moschatellina*), it seems that a period of cold weather is essential before germination can take place; others may require two seasons of winter chilling, whilst there are some that respond only to fluctuating temperatures, which may account for their fickle germination in some years. Oxygen requirements vary widely; some water plants can make a start in very low oxygen concentrations. Light can be an important factor correlated with habitat; species colonizing open ground commonly requiring light for their germination (*Epilobium* spp.). Apparently its action is to destroy the germination inhibitors, and in some cases it has been shown that a remarkably small amount of light is sufficient to do this. It is stated that one second's exposure to light of 730 candle-power, at 30° C. will raise the germination rate of purple loosestrife (*Lythrum salicaria*) from about 2 to 30 per cent. Again, other seeds are light-shy, perhaps due to photo-sensitization of germination inhibitors (*Veronica persica*). But the situation is confused, making it impossible to generalize, because these reactions to light are so often profoundly affected by temperature and the previous history of the seeds.

Chemical substances dissolved in the soil solution constitute another factor of ecological significance, as we have already seen in the response of thrift seeds to varying salinity. An interesting and highly specialized adaptation is the germination of seeds of parasitic broomrapes (*Orobanche spp.*) *only* when stimulated by chemical secretions from the roots of the right host plants. In the absence of suitable host plants the seeds may remain dormant in the ground for years. An allied phenomenon is the recent discovery by Professor Osvald in Sweden that root exudates from certain species of grasses (*Agropyron repens, Festuca rubra*) suppress the germination of various weed and crop seeds, while some *seeds* will prevent the germination of other species lying close to them in the soil (e.g. *Lolium perenne* seeds suppress the germination of *Anthemis arvensis* and *Matricaria inodora*). Clearly, the close competition of grassland makes use of chemical warfare as a means of aggression!

10

Nature of Plant Communities

INTRODUCTION

In a book on plant ecology the reader may question why only
three out of twelve chapters are devoted to plant communities:
a more or less detailed account of types of British vegetation
might have been expected to follow at this stage. There are three
main reasons why this approach has not been adopted.

In the first place, admirable descriptions of British vegetation
are already available (Tansley, 1939 and 1949, and many papers
in the earlier numbers of the *Journal of Ecology*). These are the
outcome of years of patient field work, and there seems little point
in trying to reproduce attenuated versions of them here.

Secondly, the student is far more likely to gain some real insight
into ecological problems by trying to build up an understanding
of his local plant communities from *first-hand* study, however inadequate. One of the principal aims of this book is that it should
provide enough background and guidance for him to make a start
on this. The first essentials for the background lie in an appreciation
of the behaviour of *individual* plants and their reactions to varying
factors of the environment; hence the length of Parts I and II.

Thirdly, ecology is a branch of biology which is 'growing up',
and passing from the purely descriptive, qualitative stage to one of
objective, quantitative analysis. The solution of many ecological
problems nowadays involves a mathematical treatment of numeri-
cal results collected in the field. This does not mean that descriptive, qualitative work is to be decried: it must always remain an
essential preliminary to more detailed field studies, and an important ingredient in the many aspects of ecology that do not lend
themselves to mathematical treatment. It was thought desirable

to make some attempt at introducing the student to the quantitative side, because of its rising importance. Most of the work is far
beyond the scope of this book, so one cannot hope to do more than
explain some of the simpler ways in which quantitative methods
can be applied. For further details the reader is referred to Greig-
Smith (1957) in which the whole field of quantitative plant
ecology is admirably surveyed, also Kershaw (1965), which gives a
more practical approach. The author wishes to acknowledge
his gratitude to Mr. Greig-Smith for allowing him to draw so
freely on his work in the account given in Chapter 11.

Our first concern is with the question: how does *order* come to
a plant community, making it more than just a collection of plants?
Any plant community must have some kind of history; it must
have had a beginning, from which some course of development
has been traced, perhaps over a great length of time. From
theoretical considerations we can distinguish three main factors
contributing to the final result.

(1) CHANCE

The first is chance. It may be questioned whether it is legitimate
to regard chance as a factor, but its importance is so often overlooked that it seems best to do so. It operates principally in
affecting the availability of colonizing or invading material and the
order in time in which it arrives. Obviously the species composing
a plant community must be drawn from whatever material happens
to be available, and, with the limited range of dispersal of many
species, this may be greatly restricted. If we consider the nature
of this restriction we can see that chance can play a part in several
ways: (*a*) *Chance and time.* Clearly, if an area is available for
colonization, the spores and seeds of colonizing plants will not all
arrive together, but the first few colonizers, which will not exploit
all the possibilities of the habitat, will be followed by other species
in course of time. Chance will play an important part here, in the
'choice' of which species arrive first. This is illustrated by a study
made by Godwin (1923) of the floras of a number of isolated
ponds of known ages, all within a mile or two of each other and in
comparable surroundings. The number of species was found to
increase with the age of the pond, at first rapidly then more and
more gradually. The species found growing in the heads of pol-

larded willows of different ages are said to show a similar increase with time. Assuming that a fair variety of colonizing material is available, the falling off of the rate of increase in the number of species as time goes on is doubtless an indication that the opportunities of the habitat are being more fully exploited, and there is competition between species. Suppose two species A and B of equal competitive capacity are concerned, then if A happens to establish itself first in a part of the area, it cannot be ousted by B, while B remains in possession of that part of the ground where it chanced to get in before A. We see, then, that the mosaic of vegetation, such as that forming the field layer in an oakwood, can owe a good deal to chance, though, as a rule, small differences in the local environment affecting competition between species are doubtless involved as well.

(b) *Chance and place.* The richness of variety of plant communities in comparable habitats must often differ simply because one of them happens to be more favourably situated for supplies of colonizing material. The flora of Ireland provides a good example: it comprises only about three-quarters of the number of species found in Great Britain. While some species (particularly from eastern England) may not have gained a foothold there due to climatic restriction, there seems little doubt that the factor which has excluded most of the absent species is the accident of the isolation of Ireland by the Irish Channel during the plant migrations following the Ice Age. In the same way, any area isolated by ecological barriers will suffer some measure of restriction in the material from which plant communities can be built up. There is a connection here with the depression of the tree line mentioned in Chapter 6. Following destruction of the forests on the uplands by early man, leaching has given rise to moorland soils which will no longer support oak trees. Isolated patches of soil still capable of supporting the growth of oaks may still remain at high altitudes, for example those flushed with base-rich waters; but even if such do exist, the chances of acorns reaching them are just about nil.

(c) *Chance and dispersal.* We have already discussed dispersal as an aspect of aggression in Chapter 9, and little need be added here. Clearly, seeds are not distributed entirely at random, for there is a high probability that they will fall within a particular range of the parent plant, but the exact spot must be determined to a large extent by chance. This may be important in deciding their survival, for in many areas the ground may be regarded as a patchwork of micro-habitats, some suitable, others unsuitable for the germination and establishment of seeds. Again, the element of chance in the exact position of colonizing seedlings can result in a mosaic of competing species, more or less independent of small local variations in habitat. Normally, of course there will be a greater likelihood of colonization by species with a high output of seeds and wide dispersal, like the 'opportunist' species, foxglove and rosebay willowherb.

(2) SELECTION FROM MATERIAL AVAILABLE IN THE ENVIRONMENT

The second factor contributing to bring order in our plant community is that the environment selects from the material available, that is, only those seeds germinate and establish themselves which have a range of tolerance corresponding to the environment of the habitat. We must remember that the range of tolerance at the seedling stage will generally be much narrower than that of the mature plant, so that even in comparatively favourable habitats much of the potential colonizing material may fail. It is obvious that there will be wide variation in the *degree* of selection exerted in different habitats, and in extreme cases like mountain rock-ledges, sand dunes and moorlands only a relatively small number of specialized species will have a tolerance range allowing them to survive.

So far, then we have a steady influx of colonizing material which either fails or establishes itself according to whether it can tolerate the rigours of the habitat. There is no settled interdependence between the members of the community, indeed it is not really a community at all, but just a collection of plants which have, as it were, passed an examination set by the environment. There may be physiographic changes going on, such as erosion, leaching or silting up, which would bring about gradual changes in the vegetation, but they would do it simply by altering the examination syllabus and imposing different conditions for selecting from the colonizing material. It is only with the operation of the third factor that a real community begins to emerge.

(3) MODIFICATION OF ENVIRONMENT BY PLANTS

Plants modify the environment. It must not be imagined from what has been said above that this third factor plays no part until the final stages in the process. It operates to some degree from the very beginning of the history of a community, but assumes much greater importance as time goes on. The effects are so diverse that they are best considered under separate sections.

(*a*) *Soil formation.* We have seen in Chapters 2 and 3 that a mere aggregation of mineral particles does not constitute a soil. A few specialized, hardy species (for example, lichens) can wrest a living from such a substrate, and their remains start the gradual process of building up an organic content which will eventually transform it into a fertile soil. Not only is the content of mild humus (mull) important for the supply of mineral salts, but, through its effect on the crumb structure it improves both the aeration and water-holding capacity of the soil. In the early stages of soil building it is the soil environment, much rather than that above the ground, which is gradually modified by the plants themselves so that it can support a progressively richer vegetation.

(*b*) *Layering and Dominance.* As soon as the soil has become sufficiently fertile to support the growth of taller species, layering (or stratification) is likely to develop in the vegetation. The taller plants will oust their light-loving predecessors by the shade which they cast; but if they are not making *full* use of the available light, the way is open for colonization of the ground beneath their foliage by shade-loving species. Once the taller species are growing in a fairly close stand the rest of the vegetation is dependent upon them, for they determine the microclimate in which the subordinate species grow. In effect, they produce a microclimate which, by the limitations which it imposes, greatly reduces the element of chance in what species can grow beneath them, or, in the terms of our previous analogy, *they* set the examination for the subordinate species. They are said to be the dominant species in the community. In this we see the major factor contributing to the development of a definite structure in the community.

Layering structure can vary widely in different communities. The *tree canopy* itself may be layered, indeed, it tends to become very complex in tropical forests where there are normally many different species of tree growing together (Fig. 11.2). Most British woodlands have few dominant species of tree, so that the tree canopy usually forms a single layer, though there may be a secondary layer formed by smaller trees, as in the case of yews growing beneath beeches or hazel beneath oaks (Fig. 4.2). If the tree canopy is fairly open, there may be enough light to allow the growth of a well-defined *shrub layer*, and, as a rule, little else can grow beneath this (Fig. 4.6). In most woodlands the *field layer* of herbaceous species such as bluebells, wood anemones, etc., is the most prolific both in variety and in the number of individuals. Finally, there may be a *ground layer* of mosses and small creeping herbs. One does not expect to find *all* the layers well developed in any single community, and one of the main differences in structure lies in the relative importance of the different layers. Beeches (Fig. 4.3) usually suppress both the shrub and field layers; in many oakwoods the shrub layer is poorly developed, but the field layer so rich that little light is left to support any ground layer; and in open birchwoods or ash-woods the shrub layer may be so prolific that the field layer remains sparse (Fig. 4.6). Although layering is of course most pronounced in woodlands it has its beginnings in more open communities, and may be quite well defined in examples such as heather moor.

In describing plant communities we should be alive to the fact that the term '*dominance*' is used with rather different shades of meaning. If we speak of common oak (*Quercus robur*) as the dominant species in an oakwood, the meaning is clear. The oaks form practically the whole of the tree canopy and so dominate the species of lower layers by determining the microclimate in which they grow. In describing the same oakwoods we can also say that bluebells are dominant *in the field layer* (this qualification must not be omitted), meaning that they cover the ground to the exclusion of other species. Their success may be partly due to the shading out of seedlings of rival species, but their influence over other species in the field layer is not the same as that of the oaks. In open field vegetation the degree of dominance exerted by the leading species must be less than in layered communities, and there is sometimes a danger of confusing dominance with frequency, if the ground is not fully colonized: we may sometimes label a species as dominant when we should have called it the 'predominant' species.

So far in this section, we have been chiefly concerned with the ways in which dominant species in a community determine the microclimate for the smaller plants, including their own seedlings, that is, the part which the bigger plants play in 'setting the examination' for their smaller fellows. This is a static relationship. But one of the most important concepts to be appreciated in the study of vegetation is its *dynamic* nature, arising from the fact that the 'examination syllabus' is being changed constantly through the influence of different members of the plant community, especially the dominant species.

(*c*) *Cyclic change and Pattern.* Quite obviously, the micro-climates beneath different-aged trees of the same species will show considerable differences, which are reflected in the composition of the ground vegetation growing beneath them. As any tree grows from the sapling stage to maturity, old age, and eventually meets its death, making way for new saplings, we should expect to find a progressive change in the ground vegetation recurring with each generation. Such cyclic changes have in fact been demonstrated in A. S. Watt's work on beechwoods (Watt, 1925 and 1947). He showed that bare ground beneath the close-growing saplings supports a community of wood sorrel when the trees are some 60–80 years old and the canopy opens out some-what; after a further 10–20 years the wood sorrel is succeeded by brambles. On the death of the old trees there follows a 'gap phase' during which light-loving species temporarily gain a foot-hold, and, with regeneration, the cycle begins once more. In an all-aged woodland (assuming uniform habitat) the ground vegetation would tend to show a pattern made up of a mosaic of different phases in the cycle described. The distribution of the pattern may be much influenced by the effect of catastrophes such as fire, storms, epidemic diseases or drought killing off a number of trees at one time and so bringing large areas into phase in the cycle.

Commonly the changes go further than affecting only the ground vegetation, for the dominant may produce a micro-climate in which its own seedlings do not succeed, thus paving the way for invasion by other species, which assume dominance. Changes of this kind are more often progressive, carrying the vegetation to a later stage in a succession, from which there is no return. However, Watt has shown that a phase with a different dominant can be part of a recurring cycle of changes. In some beechwoods regeneration is inhibited by the parent trees, which are eventually replaced by ash, oak or birch, but these in turn provide conditions which *promote* regeneration of the beech trees and are therefore to be seen as a phase in a cycle of changes, or **regeneration complex**. In such woods, patches of ash, oak or birch are an integral part of the *beechwood* community.

Cyclic changes of this kind are not confined to woodlands and their ground vegetation. Watt (1947) refers to examples from hummocky grassland, bracken invading grass heath, and mountain communities of dwarf heather (*Calluna*) and of mosses. They inevitably give rise to a definite pattern in the vegetation, but often this can only be detected by careful study of the communities. One of the aims of statistical methods of analysing vegetation is to provide a sensitive means of detecting pattern.

(*d*) *Succession*. Successional changes are a far more obvious feature of vegetation than the cyclic changes just discussed, and they play a larger part in our interpretation of plant communities. The principle underlying them has been implicit in much of the discussion throughout this book: it is that the changes are a product of the plants themselves. The plant community existing at any one time modifies the environment (both the soil and the microclimate above it) in such a way as to make it more suitable for other species than those dominant at the moment. These other species that come in gradually gain ascendancy and become the new dominants. The subordinate vegetation and the dependent animal communities must in turn become adjusted to the changes in microclimate which follow the new dominants.

As a community becomes more complex with time, we have seen that the dominant species play a progressively more important part in determining its structure. It is scarcely surprising, then, to find that successions, starting from a wide variety of pioneer communities in different habitats, tend to converge and follow broadly in one of a few well-defined courses. Our understanding of vegetation is greatly helped if we can recognize a community as a stage in a particular succession or **sere**. We can then form ideas about its past history (which may be confirmed by the evidence of peat

G

deposits or other remains), about the direction in which it is likely to change in the future, and how the course of these changes can be deflected by human interference. Such information may be very important in applied ecology, where land utilization is under consideration.

Of course, the progress of a sere is often so slow that the sequence of changes must be partly built on hypothesis; by putting together in order the different communities regarded as stages in the sere, like using a number of snapshots to reconstruct the sequence in a square dance. But there is usually much other evidence to support the hypothesis. Small scale-changes, can sometimes be seen within the time that accurate records are available, the direction of these changes according with the expected sequence of the sere. For example, in a succession on bare rock, islands of vegetation considered to be at a slightly more advanced stage than the rest gradually increase their territory: ponds in which silting up is hastened by the activity of a muddy stream show in a few years the vegetational changes we should otherwise expect in many generations. On slopes with a favourable south aspect the sere may have progressed further than on the more exposed north-facing slopes (Fig. 1.1). There are commonly some areas like this in seral communities where the sere tends to be held up or hastened by local environmental differences, and the vegetation of these points to the sequence that is being followed. At the edges of ponds, and in a saltmarsh succession a whole sequence of stages is set out in zonation before our eyes. Finally, the analysis of successive pollen deposits in peat or lake-bottoms can give detailed confirmation of the past history of the vegetation.

The examples given below are very much simplified, and are intended only to show the general outline of different kinds of seres. More precise details vary with different habitats, and must be worked out from field studies on each particular site. **Primary seres** (sometimes contracted to **priseres**) are those which have their origin on ground not previously occupied by vegetation. Plants can colonize a great variety of habitat, and the conditions in many demand special adaptations in the colonists if they are to survive. Broadly speaking, pioneer communities can be grouped into those adapted to tolerate either shortage of water, and made up of xerophytes (see p. 95), or an excess of it, when the members

are hydrophytes (p. 98). Extreme conditions of these two kinds
are found in the commonest sites available for colonization: bare
rock and standing water; each carries a characteristic pioneer community showing remarkable constancy regardless of climate.
Primary successions starting from these two extremes are conveniently classified as **xerarch successions** or **xeroseres**, in
which the original habitat is dry, and **hydrarch successions**,
shortened to **hydroseres**, starting from standing water.

In a typical xerarch succession on granite the first colonists are
crustose lichens, which glue themselves to the rock surface, and
form a nucleus of soil from their own organic remains and by
hastening chemical breakdown of the rock. They are followed in
due course by larger foliose lichens. During these early stages
growth of the pioneer plants can only take place when water is
available, and the little soil that is formed is liable to be washed
or blown away. Progress of the sere may be very slow indeed. The
hardy mosses like *Rhachomitrium* which come in next, speed up
the process of soil accumulation by their tufted habit of growth
trapping wind-blown particles, and they form a mat which
spreads slowly over the rock. In this mat hardy grasses make their
appearance, and the process of soil accumulation gains momentum.
In time there is enough depth of soil to support also larger plants
such as thyme and ling, and, with the advent of scrub vegetation,
pioneered by gorse, broom or wild roses, we have the beginnings
of a shade habitat on the ground. Water shortage becomes less of
a problem as the water-holding capacity of the soil is built up,
and the scrub vegetation provides some shelter from exposure.
Conditions may now have improved sufficiently to allow seedlings
of mountain ash or hawthorn to survive, and these will gradually
shade out the light-loving grasses and scrub beneath them. As the
stand becomes closer they tend to shade out their own seedlings;
regeneration fails and they are gradually replaced by more shade-
tolerant trees: birch and pine, usually followed by oak, beech or
sometimes ash. Each kind of tree will create its own particular
conditions of microclimate beneath it, and is accompanied by
characteristic ground vegetation. With the arrival of oak or beech
a state of equilibrium is reached and the community becomes
stabilized. Conditions are such that the dominant species can
regenerate themselves and keep out intruders provided there are

no changes in climate. Maintenance of soil fertility, the cycle of build up and decay, is a part of the general dynamic equilibrium which the community maintains. Such communities which terminate a succession are called **climax communities**. They are generally regarded as ecosystems in which the maximum use of the resources of the environment is achieved, the basic limitations being the climate and the parent material from which the soil was built up. Thus in any particular climate the climax communites on soil of differing origin are likely to differ somewhat in structure and in the species which compose them (that is, in their **floristics**).

In considering the succession on granite rock we have taken an extreme case. On rocks which weather more easily the early stages would proceed much faster, particularly if the rock cracks readily. We have already mentioned (p. 14) buck's-horn plantain and thrift growing directly in the cracks of slaty rocks (Fig. 2.1), and yew trees on limestone crags (p. 16). The pace of soil accumulation is speeded up locally in the cracks, and plants belonging to later stages in the succession can gain a footing earlier. Other possible sites for colonization include mountain screes, ground exposed after land-slides, and alluvial deposits on the inner banks of river bends. Successions on alluvial deposits of this kind may scarcely qualify as xeroseres unless the water level falls a great deal during the summer. On such sites the ground is comparatively fertile from the start, and soil building plays only a minor part in controlling the rate of succession. The early stages are telescoped, and small annual herbs, including many common garden weeds, make their appearance first simply because they can germinate and grow up quickly. Slower growing perennials follow crowding them out and the succession continues as in a rock succession to climax woodland.

It may at first seem surprising that a **hydrarch succession**, or **hydrosere**, starting from conditions at the opposite extreme, moves towards the *same* climax as the xerosere. Both converge to produce equitable conditions, which we call **mesic**, supporting **mesophytic** vegetation. The relationship is shown diagrammatically in Fig. 10.1.

In a hydrosere the limiting factors are the depth of the water and lack of aeration: the trend of the succession is towards the

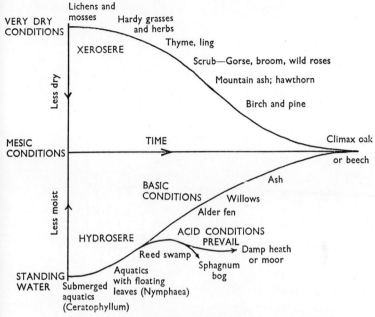

FIG. 10.1. Diagram summarizing relationships between primary
successions. (See text)

accumulation of silt, raising the soil above the water level, and
eventually lowering the water table. The first colonists grow in
open water in a pond, lake or sluggish stream. They are the submerged aquatics, usually with dissected leaves, like the various
water crowfoots (*Ranunculus* spp.), hornwort (*Ceratophyllum*) or
water millfoil (*Myriophyllum*). They are anchored in the mud
(though hornwort has no actual roots) and the maximum depth at
which they can survive must have a light intensity above their
compensation point. In water with a good deal of plankton or
other suspended matter to absorb the light this maximum depth
may not be much more than 6 ft. Any silting up of the pond at
this point will raise the level of the bottom and allow the growth
of species like water lilies for which the limiting depth is shallower.
The floating leaves shade out the submerged species and prepare
the way for the next stage in the succession as their remains help
to raise the level of the mud. Some free-floating aquatics: the
duckweeds (*Lemna* spp.) and frogbit (*Hydrocharis*) occur in the
floating leaf zone. Their only restriction to shallow water is that

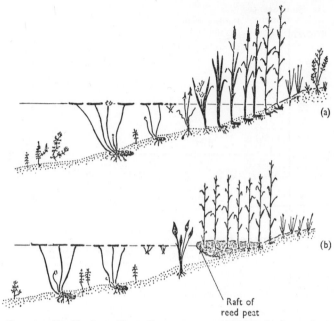

FIG. 10.2. Profile charts illustrating two successive stages in the gradual silting up of the margin of a pond, and the advance from the bank. Submerged aquatics: Canadian pondweed and hornwort. Forms with floating leaves: white water lily and frogbit. Reed-swamp zone: arrowhead, bur-reed, water plantain, bulrush (*Typha*) and reed (*Phragmites*) or reed grass (*Phalaris arundinacea*). On bank: rushes and meadowsweet

they must overwinter at the bottom of the pond, but though theoretically free-lance they seem to be kept to the margin of the pond by surface wind currents. In shallower water the horizontal spread of floating leaves is replaced by a thick upward growth of species with aerial leaves, like reeds, rushes, yellow iris and bulrushes (*Typha*). Their dense growth hastens the accumulation of silt, and the effect of their transpiration must make some contribution towards lowering the water level in the pond. This sequence may be seen as concentric zones of vegetation round the margin of any suitable pond, and is illustrated in the profile chart shown in Fig. 10.2. As such, it represents no more than a zonation correlated with increasing depth of the water, but if the succession progresses these zones migrate bodily out into deeper water and the banks of the pond close in behind them. Under certain con-

FIG. 10.3a. Fenland ditch — Spring. Note floating aquatics, and marsh marigold (*Caltha palustris*) growing on the bank

FIG. 10.3b. Fenland Ditch—Summer. Typical reed swamp vegetation

ditions the reed-swamp community may form a floating mat which extends some way from the shore: this is well seen at the head of Esthwaite Water in the Lake District.

In the stages following the reed-swamp community the vegetation is terrestrial, at first with a high water table and poorly aerated soil. Such condtions tend to favour the development of soil acidity, and the pH of the standing water may well determine the future course of the sere. If it is sufficiently base-rich to counteract this acid tendency, then fen vegetation arises, with grasses, sedges and later alders and willows (Figs. 10.3 and 10.4). These effect a gradual lowering of the water-table, and prepare the way for ashwoods and eventually a climax of oak.

In areas where the pond or lake water is poor in bases the succession may be deflected by increasing acidity after the reed-swamp stage. *Sphagnum* moss may get in and the succession proceed no further than bog vegetation. Under drier conditions this will usually drain in time and give rise to heath vegetation with ling (*Calluna*) and bilberry (*Vaccinium*) as dominants. Although heath vegetation can give place to birches, pines and eventually climax oakwoods, the soil of these heaths is generally too acid to allow much colonization by trees, and the succession proceeds no further than the heath stage. It is in equilibrium and can therefore be considered as a form of climax vegetation, usually known as an **edaphic climax**, because it is stabilized in its existing form by local peculiarities of the soil.

The vegetation in this country has been so much influenced by human activity that, apart from the early extreme stages of primary successions, few other really natural examples survive. But these pioneer communities are of great interest and are easily studied. In mountains and rocky coastal areas we can see the beginnings of xeroseres unaffected by man, but the colonization of the artificial sites which man provides makes just as fascinating a study. Roofs and old walls (Rishbeth, 1948) (*Blundell's Science Magazine*, Nos, 2 and 6), offer a rich variety of habitat, both in the chemical differences in the building stone, brick and mortar, and in the ledges and the direction in which they face. Tip-heaps in mining districts can provide good opportunities for studies in colonization (Brierley, 1956).

But before considering man's impact on natural vegetation in

Fig. 10.4. Marsh vegetation
among alders
(*Courtesy of John Markham*)

greater detail we should first look at successions in which physiographic factors play a prominent part. In those we have discussed already progress has depended largely upon biotic factors, that is, the activities of the plants themselves, but physiographic factors must always have had some influence. For example, the removal of embryo soils by wind or rain-wash must often be important in delaying the early stages of a xerosere; leaching of the soil can affect its course later on; and streams silting up a pond or reservoir will greatly accelerate the early stages of a hydrosere.

Successions in which the physiographic factors loom large are often known as **erosion successions**. Erosion may simply act as a check, setting the succession back at various stages, as in a river succession. If we follow a river down from its source we are virtually tracing the course of succession along its banks. As a young stream near its source it is cutting away its banks so rapidly that soil movement delays effective colonization; but as soon as the steep slopes become stabilized rich woodland develop in the sheltered ravine. This is very often seen in moorland country. The sere has nearly reached a climax vegetation already, but, with widening of the valley there is a retrogression, and recolonization of the drier and less sheltered slopes of the open valley proceeds more slowly. In the final stage of the flood-plain, deposition replaces erosion and a rich climax can develop. But even here erosion still proceeds on the outside of the meanders, and a telescoped succession starts afresh in the silt deposited on the inside of these bends. Islands in mid-stream and ox-bow lakes provide further interesting details of this erosion succession.

A more limited example of an erosion succession is that seen on salt-marshes; it is known as a **halosere**. Salt-marshes are common on mud banks formed at the mouths of rivers where the flow is checked by the tides. They are often refreshingly free from human interference, and show a regular zoning of vegetation corresponding to the length of time that the different levels of the marsh are submerged at high tide. This zoning also represents successive stages in the sere which proceeds as the salt-marsh gradually encroaches on the open water. The raising of the soil level is brought about by the vegetation trapping silt carried with the ebb and flow of the tide. In this it resembles an ordinary hydrosere, but the colonizing plants have two other problems to face: the instability of the mud, and high salinity. The sea-water left after inundation at high tide has a salt content of about 3 per cent, but after evaporation on exposure at low tide the soil solution near the surface may contain as much as 6 per cent or 7 per cent salt. These fluctuations and high concentration raise problems of osmotic pressure which only plants specially adapted to live in a salt-marsh **(halophytes)** can face. The upper levels of the marsh are covered only by the highest spring tides so that most of the salt can be washed out of the ground by rain. It is this progressive reduction of salinity as the level of the soil is raised which governs the progress of the succession.

Colonization of the lowest mud banks, just exposed at low tide, is begun by small green algae (especially *Vaucheria*), or sometimes by communities of eel-grass (*Zostera*). These achieve some stabilization of the mud and tend to increase the rate of silting up. At slightly higher levels glasswort (*Salicornia*) forms an open community, still submerged for much of the day. Being annuals, these plants have only a limited mud-binding effect, but they seem to prepare the way for other colonists, particularly sea manna-grass (*Puccinellia maritima*) and sea aster (*Aster tripolium*) which stabilize small islands of the mud. The flow of tidal water thus becomes restricted to the passages between these, and tends to cut quite deep creeks. There is commonly a dense growth of sea-purslane (*Halimione portulacoides*) along the banks of these creeks. The stabilized 'high marsh' comes to bear a close mixed community in which the prominent members are usually sea aster, sea manna-grass, thrift (*Armeria maritima*), sea lavender (*Limonium vulgare*), with sea-purslane still lining the creeks. Above this sea-rush (*Juncus maritima*) commonly becomes dominant, and the succession may proceed to grassland known as 'saltings' and used for grazing. The development of these is sometimes encouraged by building a wall to keep out the high spring tides. Left to itself the succession would pass to scrub and woodland.

Details of salt-marsh successions vary at different parts of the coast, and are well described by Hepburn (1952). One important variant should be noted. Wherever the vigorous rice-grass (p. 7) is present it replaces most of the stages in the succession outlined above, and may form almost a pure stand from pioneer mud-binding to the high marsh, which can be used for grazing.

Another erosion succession illustrated by coastal communities is the dune succession (sometimes called a **psammosere**). Just as the salt-marsh succession is a special case of a hydrosere, this is a special case of a xerosere with the added problems of the instability of the sand and the influence of salt spray. The dry, wind-blown sand is often pioneered by sea couch-grass (*Agropyron junceiforme*) but it is marram grass (*Ammophila arenaria*) that builds the dunes. This grass spreads rapidly by rhizome and has remarkable stabilizing properties. Its tufted growth effects a local check in the wind-speed so that sand is deposited around the plants forming a miniature dune. This effect is cumulative and gives rise to the line of foredunes, which may reach as much as 60 feet in height. The marram-grass can stand up well to repeated burial in moving sand, for its rhizome responds by growing rapidly upwards and producing a fresh crop of leaves. Little else can survive on the seaward side of the first line of dunes, but on the leeward side a number of different species are to be found. Nevertheless they do not succeed in stabilizing the sand permanently; the foredunes are never captured, and are often called the mobile dunes. Behind the line of foredunes a second, lower line is built up from the sand that escapes. They are relatively sheltered by the mobile dunes, and this time colonization is successful. They become the fixed dunes, bearing in time scrub vegetation which can give rise to woodland. The early flora of the fixed dunes is a rich one, including many annual garden weeds which complete their life cycle early in the spring and survive the summer droughts as seeds (winter annuals). Some mention of the problems of dune soils has been made in Chapter 3 (p. 35); a detailed account of the habitat and the plants found there is given by Salisbury (1952).

Few examples of natural climax vegetation have escaped man's activities in this country, but there are plenty of semi-natural woodlands, planted perhaps generations ago with the trees that would normally grow there, and maintaining themselves by natural regeneration. In many cases these must resemble natural climax vegetation quite closely. With the modern emphasis on quicker returns from fast-growing timber, plantations of conifers, sometimes exotic species, are more often seen. These are to be regarded inevitably as artificial communities. Man's greatest influence on vegetation has been to destroy the climax forest and arrest the development of secondary successions following deforestation. In arable farming the crop 'communities' are so artificial that they have no relevance here, but pastures are a different matter. With good grazing management permanent pastures are held indefinitely in a state of equilibrium, and can be regarded as subclimax vegetation maintained by biotic factors. True, the grazing factor is directly controlled by man, but it is not fundamentally different from the chalk grasslands held at the subclimax stage by rabbits. Any factor which operates continuously or repeatedly to hold up the progress of a succession results in **subclimax** vegetation. Regular coppicing or burning are examples. Physiographic factors may also produce the same effect, as when the current in a river cancels out the progressive silting up in the vegetation along its banks. Edaphic climax, for example, heath vegetation, is regarded as distinct because it represents the *maximum* development that the conditions of the habitat will permit; and, after all, the soil conditions are the climate in which the roots live. It should be noted that subclimax vegetation nearly always shows some difference from the stage through which the sere would pass if the limiting factor — grazing, coppicing, or fire — were absent. Seral grassland communities differ in their composition from subclimax grazed meadows, due to the effects of selective grazing on competition between the plants growing in the sward. Fire and coppicing will have similar effects on the balance between competing species. Thus, besides being checked, the succession is somewhat deflected.

This deflection is seen more clearly if, after some catastrophe like fire or deforestation, the succession is allowed to proceed. Its course is usually quite different from that of the unhindered primary succession and it usually leads to a different climax vegetation. The successions which follow such checks are called **secondary successions**. The main causes of secondary successions are forest fires, lumbering and grazing. In the last case the intensity of grazing is important, for if the population of grazing animals is low it may permit the growth of scrub and tree species that are distasteful, although the former dominant tree has not been allowed to regenerate. Much of our upland heath and rough pasture is the result of secondary succession. The original climax woodland was destroyed by our distant ancestors, but the land was left more or less to its own devices when they had cleared the lower woodlands on more productive soils. But, with the removal of tree cover, leaching was so intense in the upland soils that they have never been able to carry the climax woodland again. There is striking evidence for this in the changes in the proportion of different kinds of pollen found in peat deposits over this period. As the total tree pollen declines there is a spectacular rise in the proportion of grass pollen, and the pollen of various heathers makes its appearance. As the heathers are insect-pollinated and their output of pollen is therefore small compared with trees or grasses, the appearance of even small quantities in the pollen-spectrum is very significant (Fig. 10.5).

It will be seen that our understanding of vegetation is broadened if we are able to recognize plant communities as fitting into the

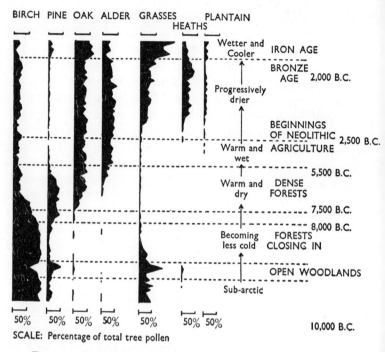

BIRCH PINE OAK ALDER GRASSES PLANTAIN

FIG. 10.5. Pollen diagram, Hockham Mere, 1940 (after Godwin, 1951)

general scheme of plant succession. There are seral communities representing some stage in a succession as yet incomplete; climax communities which may differ in detail with variations in rainfall or altitude (climatic climax) or with local soil peculiarities (edaphic climax); subclimax communities where the succession is held up by grazing, coppicing or regular burning; and finally the various stages in secondary successions. To this may be added the large area of the country under artificial 'communities'— arable crops and plantations. Some notes on climax types of vegetation in Great Britain are included in Chapter 11.

II

Aims and Methods of Studying Vegetation

Most ecological studies are directed towards gaining an understanding of the structure and interrelationships within particular plant communities; seeking an answer to such questions as how the communities have come about and what factors maintain them in equilibrium or cause them to change; whether they can be recognized as a stage in a sere; how the vegetation is affected by the varying intensities of different factors.

On the applied side, such studies of plant communities (and the animal communities dependent upon them) are closely linked with problems of land utilization and long-term conservation of the world's natural resources to meet the needs of a rapidly increasing human population. For centuries there has been exploitation of natural resources with little thought for the future. The price paid for ill-considered steps in forest clearance and agriculture is to be seen in terms of soil erosion, flooding, advance of deserts, fungus and insect plagues of crops; all due to a shifting of equilibrium in the ecological balance of plant and animal communities. In many cases these could have been foreseen and prevented had there been better understanding in the right quarters.

Even in Great Britain, as we have seen, wide areas of moorland represent the result of felling and grazing out of forests from the Iron Age onwards; and the lowering of the water-table is likely to become a widespread problem. Land utilization, forest and pasture management, the control of fungus diseases and insect pests of crops are all aspects of applied ecology, once they pass from being the 'art' of farming or forestry to a proper scientific basis; and the information upon which practice is based derives from the results of intensive studies of particular communities. Clearly, any such study would be largely barren without a description of the community concerned to serve as a point of reference which will allow repetition of the work to check results, and comparison with other communities. It would be like carrying out a number of physiological experiments with an unnamed plant. Some method of *describing* plant communities is, then, the first essential in the study of vegetation.

Description of particular concrete examples of plant communities (or **stands**) will enable comparisons to be made. If a plant community is a definite entity, as we have claimed, we should be able to find others like it in comparable habitats. They would scarcely be identical; but by comparing similar stands we might expect to find that they could be *classified* into abstract units, just as individual plants, all of them slightly different, can be classified into species. The wide variety and complexity of vegetation make some scheme of classification necessary not only for purposes of reference, but also as a means of helping the mind to grasp the material with which it has to deal.

Plant ecologists on the Continent, particularly those of the Zurich-Montpellier school of thought, have been mainly interested in studying the classification of vegetation for its own sake, and they seek to perfect a system in which plant communities can be classified into categories resembling the familiar species, genera, families and natural orders of flowering plants. They are therefore much concerned with describing and comparing numerous stands of vegetation (concrete examples), and thus deriving abstract ideas of plant community 'species', 'genera' and higher categories. The term **phytosociology** is commonly used to describe this aspect of plant ecology, though it is not really a correct use of the word which was coined by Continental workers to include all aspects of the study of plant communities.

In Great Britain interest has centred rather on successional relationships and intensive studies of particular plant communities and their habitats; general classification of vegetation has remained largely descriptive. Where Continental workers have been concerned with analysis of vegetation on a broad scale, we have worked more on the detailed analysis of smaller units. With this divergence of aims it is scarcely surprising that our methods of study and criteria used to characterize plant communities are different. As the techniques of phytosociology can have little application in school work, no details will be given here; but for those who wish to know more about them, they are well summarized by Oosting (1956) and a detailed discussion of their value is given by Poore (1955–6).

Thus, while not much progress has been made in Britain towards perfecting any broad scheme of classification, British vegetation has been described with a thoroughness probably unmatched in any other country (Tansley, 1939), and different communities can be recognized and named, so that work can go ahead on intensive studies.

CLASSIFICATION AND DESCRIPTION

The general trend of plant successions, whatever their origin, is towards a characteristic climax type, the nature of which is determined largely by climate over large areas. The different climax types, or **plant formations** are the main natural vegetation types of the world, like tundra, steppes and prairies, coniferous forest, deciduous summer forest or evergreen tropical rain forest. They are easily recognizable, as their common names show, by the distinct growth forms of the dominant species. The climax vegetation of these plant formations is, of course, not uniform, but made up of a number of fairly well-defined smaller units which we call **associations**. They are characterized by the *particular* species which dominate them: if there is a single dominant (for example, in a beechwood) the community is called a consociation, if more than one dominant species (for example, ash-oak wood sometimes found on deeper limestone soils), it is an association. Particular consociations are referred to by adding the suffix *-etum* to the stem of the generic name of the dominant species, for example, a beech consociation is a *Fagetum*. If there is any risk of ambiguity about the species, then the specific name is added in the genitive case, for example, *Fagetum sylvaticae*, *Quercetum roboris* or *Quercetum petraeae*. We have already seen that some ecologists abroad have very different ideas from our own about the classification of plant communities. Unfortunately the term 'association' is used in different systems to express entirely different ideas, so that in consulting wider literature the student should be alive to the risk of confusion.

Smaller distinct communities occurring locally within the consociation or association are known as **societies**. The term is elastic and can be used to refer to communities of widely differing size; from an area of ash trees growing in the damper part of an oakwood to a patch of bluebells on the ground or a group of ferns growing along the branch of one of the trees. The society is usually dominated by one species, which may be a subordinate within the framework of the consociation as a whole. It may also have a characteristic structure in its own subordinate species.

Communities of the rank of consociation or association, and societies can be recognized not only in climax vegetation, but at every stage of a succession. When it is desired to emphasize that a particular community is seral, that is, a stage in a succession, the terms **associes** or **socies** are used instead. For example, a reed-swamp community occurring as a stage in a hydrosere and dominated by common reed is called a *Phragmites* consocies.

The different communities of association rank within a climax formation may be the result of variation in climate (especially the rainfall/evaporation ratio), local soil conditions, or historic factors (for example, glaciation, isolation by sea) affecting the availability of seed-parents. The following very brief notes outlining some of the main types of climax vegetation in Great Britain will serve to illustrate this. Detailed descriptions are available in Tansley (1939).

(1) *Quercetum roboris* — Oakwoods of common or pedunculate oak (*Quercus robur*) (Fig. 4.4). These occur especially on clays and damper loams in the midlands, south and eastern England. The field layer in these woods is well developed and includes a wide variety of different species; dog's mercury, wood sanicle and bluebells are among the most conspicuous members.

(2) *Quercetum petraeae* — Oakwoods of durmast or sessile oak (*Q. petraea*). These are characteristic of rather poor siliceous soils of the north and west, and are commonly found on steep slopes, as in the Welsh hills and the Pennines. The field layer is typically much poorer in species than that of the lowland pedunculate oakwoods, with bracken, woodrush (*Luzula sylvatica*) and various grasses much in evidence.

These two species of oak are readily distinguished by the differences in their leaves (see Fig. 11.1), and by the fact that in *Q. robur* the female flowers, and later the acorns, are stalked (pedunculate), while those of *Q. petraea* are sessile. When growing together they hybridize freely, and lowland oakwoods on rather acid soils often consist almost entirely of hybrid trees showing a mixture of these diagnostic characters.

(3) *Fagetum* — Beechwoods (Fig. 4.3*a* and *b*) Characteristic of chalk downs and the soft oolite limestone of the Cotswolds. It is possible that the historic factor may have played a part in their distribution, in that seed dispersal in beech is so poor compared

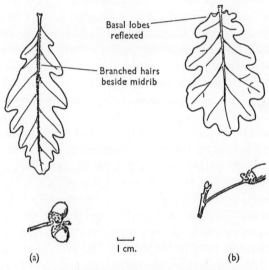

FIG. 11.1. Acorns and the under-surface of leaves of (*a*) dur-mast oak (*Quercus petraea*) and (*b*) common oak (*Q. robur*)

with ash, its chief competitor on limestone soils. It has been suggested that the prevalence of ash in the north and west may be connected with its faster migration when temperate vegetation returned to Britain after the Ice Age. But beech cannot tolerate wet soils; indeed its occurrence on chalk is probably more favoured by good drainage than by the chemical nature of the rock.

The field layer in beechwoods is typically very sparse indeed. Among the species most often found are dog's mercury, wood sorrel, wood sanicle and wood barley-grass (*Hordelymus europaeus*), but they rarely form a dense carpet unless the wood is open in character. Yews may occur forming a secondary tree layer which suppresses the ground vegetation entirely.

(4) *Fraxinetum* — Ashwoods. These may represent a seral stage (an associes) to be replaced ultimately by climax beechwood; but in the wetter hills of the north and west, especially on the hard Mountain Limestone (Lower Carboniferous), ashwoods are themselves the climax vegetation. The open leaf canopy favours a rich field layer, often in nearly pure stands of a single species: dog's mercury, ivy, ground ivy (*Glechoma hederacea*) or melic-grass (*Melica nutans*).

(5) *Moorland and Heaths.* These include a number of communities where the normal climax vegetation has not been able to develop owing to local limitations of climate (usually altitude) or of soil. It is often difficult to make a clear separation of the climatic and soil influences, for the same climatic extremes which check the succession also favour podsolization of the soil.

Typical upland moors, as those of the Pennines and the Scottish Highlands, may be regarded as climatic climax. They are subject to high rainfall/evaporation, low temperatures and exposure; conditions which hold up the decay of organic matter in the soil and result in the formation of a deep layer of peat. The vegetation varies with the nature of the parent soil material and with drainage. Broadly speaking three main types can be recognized: grass moor (especially in Westmorland, Yorkshire and parts of Wales), cotton-grass moor (common in the Pennines and the Scottish Highlands) and heather moor (widely distributed). The grass moors are usually dominated by purple moor-grass (*Molinia caerulea*) which is very tolerant of acid, wet soils: where the ground is better drained, mat-grass (*Nardus stricta*) often replaces it. Cotton-grass (*Eriophorum vaginatum*) covers large areas where the soil is waterlogged. Heather moors are usually dominated by ling (*Calluna vulgaris*) often co-dominant with bilberry (*Vaccinium myrtilis*), giving place to cross-leaved heather (*Erica tetralix*) and various bog plants, like bog asphodel (*Narthecium ossifragum*) and the sundews (*Drosera* spp.) growing in sphagnum moss in the wetter places.

Typical heaths are lowland edaphic climaxes; the results of secondary successions following deforestation on poor sandy soils and gravels. The soil is apt to be dry, with only a thin layer of peat, though waterlogged pockets are common where drainage is impeded by the formation of a hardpan 2–3 ft. below the surface.

On the drier areas bracken is often dominant; but heathers, especially bell-heather (*Erica cinerea*) sometimes mixed with bilberry are common, also gorse or broom. Various species of bent-grass (*Agrostis*), milkwort (*Polygala serpyllifolia*), tormentil (*Potentilla erecta*) and the lichen *Cladonia* grow as a secondary layer among the heather. The damper pockets have typical bog vegetation: sphagnum with bog asphodel, sundews, butterwort (*Pinguicula lusitanica*) and the southern species of cotton-grass (*Eriophorum angustifolium*), clumps of bog myrtle (*Myrica gale*); and cross-leaved heather bordering on the drier ground.

Many examples of heathland are not so simply explained as this. To the west, where the rainfall/evaporation ratio is higher, climate plays a more important part in their origin, especially on moderately high ground. The distinction between damp heath and moorland tends to become artificial in such cases. At the other end of the scale, where soil conditions are not extreme, pine and birch may colonize the ground giving rise to open pine-heath association seen in some of the commons around London. In other places this invasion by trees may be checked by frequent heath fires, and the heath vegetation is to be regarded as subclimax.

From the outline classification it will be seen that the unit of vegetation most requiring detailed description is the association or consociation. In attempting to describe a community of this rank we should obviously begin by naming the dominant(s), and listing any other species which achieved local dominance. If the dominant species are trees, then estimates should be made of the height and age of the larger specimens, their average spacing, also notes about regeneration and the proportion of younger trees coming on. A general idea of the layering in the community could then be given, explaining whether the shrub layer, field layer and ground layer were sparse or well-developed, and listing the dominant species or those forming prominent societies in each layer. Where particular societies show obvious correlation with local variations in the habitat, these should be noted.

Though this would give a general picture of the structure of the community the information is not in the most convenient form, and may not be adequate, for making detailed comparisons between different stands of the same association. The next step is to make our species list as complete as possible, and give some indication of the relative abundance of each species. Assessment of frequency made by inspection (in terms of 'dominant', 'abundant', 'frequent', 'occasional' or 'rare') leaves room for large personal errors, especially in comparisons made at different times of the year, or by different workers. Accuracy can be improved by applying quantitative methods, but the measure of frequency has still its limitations. The use of this, and other measures on a quantitative basis is discussed later in this chapter. More reliable subjective estimates can be made using a system proposed by Braun-Blanquet (Kershaw, 1965). Two scales are used, one taking into account the number of individuals and cover of the species, the other its 'sociability' or grouping into clumps, patches etc.:

+ = sparsely or very sparsely present, cover very small

1 = plentiful but of small cover value

2 = very numerous, or covering at least 1/20 of the area

3 = any number of individuals covering $\frac{1}{4}$ to $\frac{1}{2}$ of the area

4 = any number of individuals covering $\frac{1}{2}$ to $\frac{3}{4}$ of the area

5 = covering more than $\frac{3}{4}$ of the area

Soc. 1 = growing singly, isolated individuals

Soc. 2 = grouped or tufted

Soc. 3 = in small patches or cushions

Soc. 4 = in small colonies, in extensive patches or forming carpets

Soc. 5 = In pure populations

An entirely different approach to the problem of describing plant communities can be made by assessing the proportions in which different life forms are represented in the vegetation (see Chapter 5). This cannot give a complete description in itself, but the additional information may be valuable in particular studies, especially those connected with seral communities.

QUANTITATIVE METHODS FOR INTENSIVE STUDIES

Comparison of closely similar stands, or the detection of variation or change in a single stand may require more accurate description calling for quantitative methods. We cannot measure the whole stand, so we must adopt some form of sampling and rely on measurements taken from a number of small sample plots. Statistical analysis can be a valuable tool, giving a means of assessing the degree of probability that the values obtained from samples are, with any desired accuracy, the same or nearly the same as those which would be obtained by measuring the whole population of the stand.

A word or two of warning may be needed here. First, it should be remembered that statistical analysis does not *prove* a point; it only tells us the probability of the same result being achieved by chance. The odd, unexpected chance may always turn up in sampling — in germination tests on seeds one *might*, by a freak of chance choose in a sample the only seeds in the whole sack that were viable. Secondly, statistical analysis does not in any way reduce the need for the faculty for sound personal judgement which comes as a result of keen observation and wide experience in the field. It is easy to become so engrossed in figures and calculations that one misses what a good field naturalist would see at once. In any event a most careful preliminary survey is always needed before the details of intensive studies can be planned. This brings us to the third point: any investigation which is to be subject to statistical analysis must be carefully planned to ensure that the sampling method and the kind of data collected are in accordance with the statistical tests to be applied. For example, most of the tests applied are based on the mathematical assumption that the samples are taken from one more or less normally distributed population, and may require that the samples are taken at random. If neither of these conditions is satisfied a considerable amount of work may be expended in collecting data and going through laborious calculations to reach conclusions that are quite valueless.

SAMPLING

In sampling we take small parts of the area being studied, and make intensive observations on those. Three kinds of sampling procedure are in common use:

(a) *The Line Transect*. This is a rapid but insensitive technique, particularly suitable for investigating transitions in vegetation due to some factor showing a gradient, such as height of water-table, or salinity. A line is taken arbitrarily running across the zoning of the vegetation (and hence in the direction of the gradient); it is marked in some way, for example, by a piece of string or a surveyor's chain, and recordings are taken along its length. The simplest form of recording is to note the species in contact with the line at fixed intervals, together with data about any habitat factors that are being measured. The results may be summarized conveniently in a profile chart (Fig. 10.2). A variant of this is the *bisect* (Fig. 11.2) in which differences in layering in woodland samples may be shown diagrammatically for comparison.

Line transects may be used to collect more precise information, such as an assessment of percentage cover by different species, for example, plantains on a lawn, by recording the length of line covering the species concerned in a large number of parallel tran-

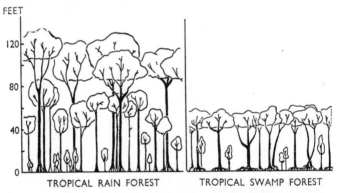

FIG. 11.2.   Bisect showing in diagram form the layering of forest tree canopy. (After Beard, 1944 in Oosting 1956)

sects. In dense shrubby vegetation this technique can be used when it is impracticable to lay down quadrats.

(b) *Quadrats*. Quadrats are sample areas, traditionally square, as the name implies, though not necessarily so. Circles have been used, and it has been shown that in many cases rectangles can give more information than squares of the same area. A quadrat usually takes the form of a light wooden frame which is set down at various spots in the area being studied, for the collection of detailed information. Alternatively, the quadrat may be pegged out with meat skewers joined by pieces of string. The *kind of information* to be collected, whether frequency, density, or percentage cover by different species will depend on the problem to be solved, and these measures will be discussed below. *Size of quadrat* again depends upon the type of vegetation and the problem in hand, and it is sometimes difficult to decide on the best size. This is particularly important when measuring frequency (p. 212), where the size of quadrat influences the results obtained — for example, counts of species occurring in small quadrats (say 1 ft. square) distributed at random in an oakwood would record few, if any, oaks. For other measures, such as density (number of individuals per unit area), percentage cover, or yield, the results are independent of quadrat size (if sufficient samples are taken) and the main consideration is that the quadrat should be larger than the unit under observation, whether individual plant, tussock of grass or recurrent pattern in the vegetation. Too small a quadrat increases personal errors due to 'edge effect', that is, decision whether plants actually on the edge of the quadrat should be included or not: too large a quadrat increases the labour unduly.

The second decision which must be made concerns the *number of quadrats* to be used. As might be expected, no hard and fast rules can be laid down; but it should not be too difficult to settle the matter if the following considerations are kept in mind. First of all, the accuracy of the information we obtain from our sampling can always be improved by increasing the number of quadrats. A very small number of quadrats corresponds to a small number of observations, and our total sample is less likely to be really representative of the population as a whole. In attempting to fit meagre data to a normal distribution curve this is reflected by very large values for the standard deviation and the standard error of the mean (p. 141) because the number of observations (N), is small. Any deductions that we can make about the whole population from our sampling are therefore of little value, since the population mean may deviate so widely from that of our sample if the *S.E.* is large. It follows that the number of quadrats that we use must be sufficiently large to give a relatively low figure for the standard deviation when the data are fitted to a normal distribution curve. Further, if we are sampling to find the *density* of different species (p. 216), it must be remembered that accuracy varies with the number of individuals counted rather than the area sampled, so that more quadrats are needed for sparse than for abundant species. Determinations of *frequency* (p. 212) based on small numbers of quadrats are liable to be very inaccurate indeed; at least 100 samples should be taken.

Placing of Quadrats. Here again, the choice must rest upon the kind of information being sought and the nature of the problem. Where studies of comparatively rapid changes in vegetation are being made, and there is a good chance of the ground being free from outside interference for some time, *permanent quadrats* are valuable. A typical problem where this technique can be used to advantage is the study of changes in ground vegetation following the felling of oaks and replanting with conifers. If only the changes in individual quadrats are to be studied, then the site can simply be chosen so as to give typical representation of the conditions. If, however, data are required for statistical analysis (how far sample quadrats represent the whole stand) then a number of samples distributed at random must be used. This eliminates the personal factor, which, in the choice of a 'typical' site, imposes preconceived ideas as to what is typical.

Random sampling. If an area were *really* uniform all the sample plots would yield the same information; indeed there would be no point in taking more than one sample. But there is always some variation, so that different sample plots give different information, and statistical analysis helps us to decide how far this information applies to the whole stand. The basic theory underlying the statistical analysis of the information we have collected assumes that the samples have been taken at random. This means that, if the area under study is considered as being divided up into a large number of sample plots, then all plots have an equal chance of being chosen each time the site of a sample quadrat is selected. This is important, because it means that no valid conclusions can be drawn by applying statistical techniques to samples which have not been taken at random.

It is often thought that to secure random distribution or quadrats, all one has to do is to shut one's eyes and throw the quadrat into the air a number of times, collecting information each time from the place where it falls. This will certainly reduce the bias from the human factor; but mathematical examination of the results shows that the distribution achieved by this method often departs widely from random. Probably the easiest way of obtaining random distribution is to fix two lines at right angles, at the edges of the area of study, to serve as axes, and use pairs of random numbers as co-ordinates to locate the positions of quadrats.

Approximate measurement, for example, by pacing out the distances will give sufficient accuracy. Pairs of random numbers may be obtained from tables of random numbers given in statistical textbooks. (Snedecor (1956), Fisher and Yates (1943)).

Just as, if we toss a coin a number of times, we may have a run of 'heads' five or six times in succession, and perhaps a considerably higher total of 'heads' than 'tails', so random sampling can also be capricious. It often happens that quite wide areas are left unsampled, while elsewhere several quadrats may be located close together. A more even distribution can be obtained by systematic placing of the sample quadrats, which is to be preferred if the problem is one in which statistical analysis may be of little help.

Systematic placing of Quadrats. A grid system (Fig. 6.6) is generally used to secure the most even placing of quadrats. Taking an arbitary base line *AB*, perpendicular lines are run off from it at fixed distances, and quadrats sited at regular intervals along these parallel perpendicular lines. The parallel grid lines may be fixed quite rapidly by compass sightings, and pacing may give sufficient accuracy for locating the positions of the quadrats.

This method of placing quadrats gives even spacing of samples, and does away with the human factor in choosing sites. It is a convenient technique to use when comparing vegetation or the performance of one species at different distances from the base line, when it is known that there is a gradient in one important controlling factor. For example, *AB* might represent a stretch of river bank, with the ground sloping upwards along the parallel grid lines, giving a gradient in soil depth above the water-table. The performance of bluebells in increasing shade near the edge of a wood (p. 78) was studied using this kind of layout.

An alternative arrangement of quadrats, suitable for studying zonation in vegetation, is the belt transect, which amounts to placing them in a row, touching each other. The collecting of information involves a great deal of labour unless the area covered is very small.

It will be seen at once that systematic arrangement of quadrats does not give random distribution. For example, if it is decided to place quadrats at 10-ft. intervals from the base line, then the position of *all of them* is fixed with the laying of the first quadrat, and plots 5 ft., 15 ft., etc., from the base-line can never be sampled.

Where random distribution is important for mathematical reasons, then it can be achieved by a combination of these methods: the ground to be sampled may be divided up into a number of blocks of equal area, and random samples taken within each block.

(*c*) *Point Quadrats.* This technique provides a rapid and potentially accurate method of estimating percentage cover, and is particularly suitable for use on lawns and closely grazed pastures. In principle it depends on recording the presence or absence of a species in a large number of very small quadrats. As the quadrat

FIG. 11.3. Point frame for sampling grassland vegetation

size is reduced, the proportion of quadrats *partially* covering the species becomes smaller, until at pin-point size it can safely be ignored, and every 'quadrat' scores either a 'hit' or a 'miss'. Suppose that in an area of ground 1,000 times the quadrat size 10 per cent is covered by a particular species A; then in 100 samples taken at random we should expect about 10 to fall on species A. In practice, then, all that is necessary is to let down a number of pin-points at random on to the vegetation, and record the number of 'hits' and 'misses' on the species we are considering. If the sample is large enough, the proportion of hits to the total of hits and misses together will give a reasonably accurate assessment of the percentage cover of that species. Groups of 10 pins, set 2 in. apart in a frame (Fig. 11.3) are commonly used. As the pins slide freely, the apparatus can be used on uneven ground. It is quite easy to make this apparatus, using brass curtain rods for the framework, and bicycle spokes, sharpened at the ends, for the pins.

In the actual placing of the frames to secure random sampling the same considerations apply as for the placing of quadrats just described. There are mathematical objections to using an arbitrary grouping of pins, which, when there is marked clumping of the plants being investigated (for example, patches of daisies on a lawn) may outweigh the advantages of speed in collecting data. In such cases it may be better to record for only one pin (which may be chosen at random) for each siting of the frame.

MEASURES OF VEGETATION AND TYPES OF INFORMATION THAT CAN BE COLLECTED FROM SAMPLES

In making detailed studies of sample areas, various measures are available for conveying information about the plant communities. Of these, each is suited to a different purpose, and in planning quantitative work careful consideration should be given at the start, both to the sampling technique and to the choice of a suitable measure. We must consider, then, what the different measures really tell us, and what pitfalls are associated with their use.

Species lists. One of the simplest techniques is to list species in sample quadrats or at intervals along a line transect (really a row of point quadrats). This may be used to compare two sites (for example, a control and an area from which grazing animals have been excluded), but it does not give very clear results, and yet may entail a good deal of labour if the quadrats used are not very small.

If the total number of species recorded is plotted against the *number* of quadrats investigated (or against increasing *size* of quadrat) a species/area curve is obtained (Fig. 11.4). The flattening of the curve indicates that, above a certain number of quadrats (or quadrat size), few fresh species are observed, and this has been used to give some guide in deciding on the number and size of quadrats needed for adequate sampling of a plant community. It must be emphasized, however, that the species/area curve can only indicate the number or size of quadrat suitable *for making species lists*, and has *no relevance at all to quantitative measures* such as *density or frequency of species*. Even when used in connection with species lists it can be misleading, as the conclusions drawn depend

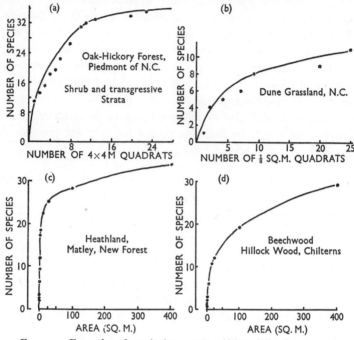

FIG. 11.4. Examples of species/area curves ((a) and (b) after Oosting 1956; (c) and (d) after Hopkins, *J. Ecol.*, 45, 2 (1957))

to a large extent on the scale used on the different axes of the graph. This difficulty can be partly overcome by locating the point on the curve where the slope corresponds to some definite arbitrary relationship, such as a 10 per cent increase in the number of species for a 10 per cent increase in area sampled.

Charting. This is of value chiefly in permanent quadrats where changes are being followed in detail. Greater accuracy can be obtained, especially in large quadrats, if a grid of stretched strings is superimposed on the vegetation, and the chart drawn on graph paper. In the United States a good deal of use has been made of pantographs for this purpose. Where there are few species, the charting of small quadrats can be used to determine percentage cover.

Frequency. An obvious preliminary way of describing vegetation is to list species as common, locally common, occasional, or rare. Because of the varying growth habit of different species, these frequency categories, when determined by inspection, are more a subjective estimate of *conspicuousness*, to which quite different variables, for example, percentage cover, density (individuals per unit area), and grouping of individuals must make a contribution in the mind of the observer. Not only, then, is this method liable to errors from the personal factor and from differences in the appearance of the vegetation at different seasons, but also the results lack precision in that they are based on simultaneous estimates of two or three different measures which may vary independently.

More precisely, the frequency of a species is measured by recording its presence or absence in a number of quadrats taken at random in the area under study, the frequency being equal to the proportion of the total number of the quadrats in which the species occurs. As frequency is defined as the chance of finding the species rooted within the sample quadrat in any one trial (sometimes called 'rooted frequency'), it is clear that the result will depend upon the size of the quadrat used, being higher, of course, for larger quadrats. When frequency (usually expressed as percentage) is used as a measure, therefore, the size of the quadrat used in finding it must always be stated. The term 'shoot frequency' is sometimes used where specimens rooted outside the quadrat area are counted as present if parts of their shoots or leaves come into the quadrat.

Raunkiaer (1934) in comparing different plant communities determined frequency percentages of all the species present, and grouped them into frequency classes, a procedure called **valence analysis;** but now rarely used.

Class *A*: Species occurring with frequencies of   1–20 per cent
   ,,   *B*:   ,,   ,,   ,,   ,,   ,, 21–40 per cent
   ,,   *C*:   ,,   ,,   ,,   ,,   ,, 41–60 per cent
   ,,   *D*:   ,,   ,,   ,,   ,,   ,, 61–80 per cent
   ,,   *E*:   ,,   ,,   ,,   ,,   ,, 81–100 per cent

It was found generally true for all communities that the percentages of species in the different frequency classes gave a frequency polygon like that shown in Fig. 11.5. This is summarized in Raunkiaer's 'Law of Frequencies,' which states that: class $A > B > C \gtrless D < E$. Class *A* includes the numerous scattered rare species with low frequencies, and class *E* will always tend to be high as it includes all the dominant species in the community. The use of these frequency diagrams as a means of comparing communities by studying slight differences in the frequency distribution proved unfruitful, as it has been shown that their form is so much affected by the method of sampling used.

Two limitations of the use of frequency as a measure should be noted. The first is that, because it is deduced from samples over

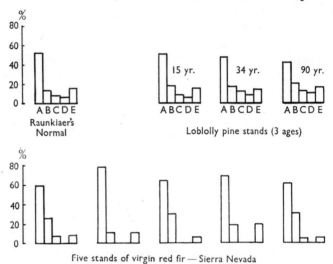

Five stands of virgin red fir — Sierra Nevada

FIG. 11.5. Frequency polygons from valence analysis of differnt stands of vegetation compared with Raunkiaer's 'normal' (based on an average of more than 8,000 frequency percentages).
The stands of virgin red fir were extremely homogeneous and similar to each other, though widely distributed along the Sierra Nevada. (After Oosting, 1956)

a comparatively wide area, it cannot be determined locally and used to examine the effects of habitat factors varying in small areas. An alternative is however available in the measure of 'local frequency', which is estimated from quadrats subdivided by a grid For each position which is sampled a percentage local frequency is obtained from the proportion of grid squares in the quadrat in which the species occurs. The results have little significance unless the quadrat is large in comparison with the size of the species under investigation. This technique has already been mentioned as one means of investigating the zonation of vegetation around a rabbit warren (114).

The second limitation is that the frequency of a species, which is obviously related to its density in the area, can be considerably influenced by clumping and pattern of distribution. Fig. 11.6 taken from Greig-Smith (1957) will make this clear. The dots represent individual plants, and each square contains the same number, so that the densities in the three squares are the same. If however, the community in (a) were sampled for frequency with a quadrat of the size shown, the finding would be 100 per cent; in (b) a much lower frequency would be found, and in (c) a value intermediate between the two. From this it will be seen that two species having the same density in a community may show widely differing percentage frequency values (when sampled with the same sized

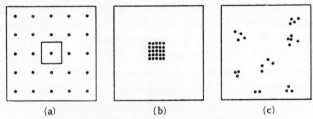

(a)   (b)   (c)

FIG. 11.6. Three different distributions having the same density. A quadrat in position in (a). (See text)

From P. Greig-Smith: Quantative Plant Ecology: Butterworth

quadrat), indicating a difference in their pattern of distribution. The analysis of pattern by sensitive mathematical techniques holds promise as a means of gaining new understanding of the nature of plant communities.

Despite its limitations and rather vague biological meaning, frequency is so easily and quickly determined that it remains a useful tool for describing plant communities.

Presence-or-absence records in a set of sample quadrats, when collected for a *number of species* simultaneously can be used to show association (or the lack of it) between different pairs of species. Data of this kind can be analysed statistically using an electronic computer, to reveal any small differences in the vegetation that might otherwise pass unnoticed. Where this promising technique has been used the differences shown up by the computer have subsequently been found to correspond to recognizable habitat differences, for example, the edge of an area which was drained many years back, or where there had been a fire (Williams & Lambert 1959).

H

Density. Density is a measure of the number of individuals per unit area. The term 'individual' may be variously interpreted to suit different problems and different kinds of plant material, for example, in working with grasses tillers (axillary shoots) or tussocks (clumps) may be counted. With some species there is no convenient unit which can be regarded as an individual, so that density cannot be used as a measure. Such plants commonly lend themselves to percentage cover determination which can be used instead.

Density values (the unit area must be stated) are not directly dependent upon the size of the sampling quadrat as in the case of frequency. Nevertheless, some care in the choice of the quadrat size will ensure greater reliability of the findings for a given amount of labour. Apart from common-sense judgement based on the average size of the species or pattern which is being studied, two considerations should be borne in mind:

(i) *Edge effect.* There is always room for some degree of personal error in deciding whether individuals occurring along the edges of the quadrat should be included or not, and this leads to differences between the results obtained by different observers. The use of small quadrats, or long, narrow rectangles, which may be convenient in some other respects, will tend to increase the error due to this factor as the ratio of edge to area is then greater.

(ii) If we intend to use statistical tests (for example, in comparing densities of a species in two populations), then the number of quadrats containing different numbers of individuals should follow approximately a normal distribution curve (Fig. 8.3) as it is on this that the statistical tests are based. In practice, for very small quadrat sizes the distribution is markedly asymetrical. A rough guide in choosing the size of the quadrat is to have it large enough to ensure that more samples contain one individual than none at all. Some system of random distribution would also be required in this case.

As the average distance between individuals in a community is inversely proportional to density, some attention has been paid to the possibilities of using this as a measure and so avoiding the need to lay down sample plots. Its most obvious application is in forestry. Details of the methods are outside our scope, but they are reviewed by Greig-Smith (1957).

Cover. This is a measure of the proportion of the area covered by any species, and is usually expressed as a percentage. It is a convenient measure in short grassland communities where determination of density would be very laborious or impossible. It may be estimated directly, or measured from charted quadrats if no great degree of accuracy is required: personal errors will come into both of these. As it is a measure independent of the sampling technique employed, quadrats of any convenient size may be used, and for short grassland communities the point quadrat technique already described (p. 210) is both rapid and accurate. In dense scrub vegetation, where the use of quadrats would be difficult, cover can be estimated reasonably well by using a number of parallel line transects. The total length of transects lying across the species in question, expressed as a percentage of the overall length of the transects gives the cover value.

It should be remembered that, except in open communities, the total cover of all species together is likely to exceed 100 per cent owing to overlap or layering of the vegetation.

Measures of Performance. Particularly when investigating the effects of soil factors, such as root competition, different levels of nutrient supply, one often requires to use some assessment of vigour or performance. The most obvious measure is yield. For this the plants are harvested, generally using small shears and the weight of material is determined (preferably dry weight). Other measures of performance, such as number of flower stalks per plant, flowers per inflorescence, length and number of runners formed, etc., will suggest themselves in particular problems, and can be used in conjunction with some simple sampling method.

TREATMENT OF DATA COLLECTED

Statistical treatment of the data collected in field studies lies outside the scope of this book, but it is hoped that some inkling has been given of the *kind* of way in which statistical analysis can be of help. It is likely to play an important part in advancing our understanding of plant communities in the future. For further details on this aspect the reader is referred to Greig-Smith (1957), Kershaw (1965), with the help of some text-book on statistical methods in biology, such as Snedecor (1950), Chambers (1955), or Mather (1951). Meanwhile, what can we do with our present knowledge to get the most out of the data that we have collected?

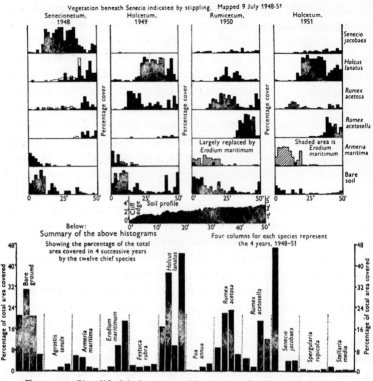

FIG. 11.7. Simplified belt transect histograms showing the changing dominants over a four-year period on the Neck, Skokholm

From Mary E. Gillham: Ecology of Pembrokeshire Islands: Journal of Ecology, 43

These data will usually be in one of three possible forms:
  (i) Information about two communities to be compared.
 (ii) Information about changes occurring in a community over a period of time.
(iii) Information about variation in vegetation which we are seeking to correlate with environmental differences.

Data from comparisons will usually take the form of frequency, density, or cover data for individual species from a number of samples. The first step is to find the mean value of the samples and examine their range of variation. A comparison may now show such obvious differences that conclusions can be drawn without recourse to any further treatment. If, however, a statistical test for significance of the difference between the means seems necessary, we can arrange the data in frequency classes, calculate the standard error of the means, and from these the standard error of the difference between the means, exactly as shown on pp. 142–146. Comparing this with the observed difference between the two means, we can judge the probability that the difference between the two sets of results reflects a true difference between the two communities rather than a chance difference arising out of the sampling of variable material.

The second type of problem may sometimes involve direct comparison of data collected by sampling the community at different times. This is a particular case of (i), and can be treated in the same way. Generally, however, we want to go further than just establishing that a change has occurred in the community, that is, that it is different from what it was — we are interested in the *direction* or trend of the change. It is here that graphical methods are often of considerable help. Fig. 11.7 illustrates changes over four successive years in a permanent belt transect of 50 ft., working inland from the cliff edge of the Island of Skokholm (Gillham, 1955). The changes are correlated with fluctuations in grazing intensity due to the varying rabbit population. In the upper half of the figure the percentage cover of the principal species present is shown by the vertical columns of the histograms, the horizontal scale representing distance from the cliff edge in feet. Each column represents the same area in four successive years. The histogram below summarizes the variations in the percentage cover of the individual species over the four years.

In the third kind of problem we have two sets of data: a series of observations or measures of the vegetation, each with a corresponding reading of some environmental factor. What we have to decide is whether the two sets are correlated. The simplest case is that of a transition where the environmental factors show a gradient, and for this kind of problem data may be collected from a line or belt transect, or from quadrats placed at intervals along transect. Again, diagrammatic representation of the data may help us to interpret the results. As an example, Fig. 11.8 shows the distribution of the principal flowering plant species growing along the banks of the estuary of the River Exe (Gillham, 1957). Although the horizontal scale is in miles from the river mouth, it could well have been a scale of decreasing salinity of the water. In this case the breadth of the columns indicates the relative

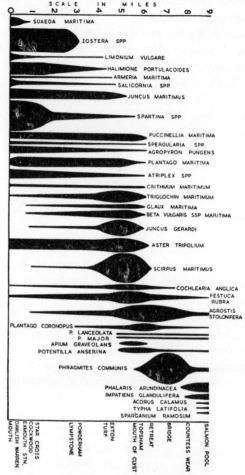

FIG. 11.8. Distribution of the principal flowering plant species growing along the banks of the Exe Estuary in relation to water salinity

From Mary E. Gillham: Vegetation of the Exe Estuary in relation to Water Salinity: Journal of Ecology, 45

abundance of the different species, but does not represent any absolute value. To depict more precise data the values for each species could be plotted one above the other as a series of histograms, against a horizontal scale representing the graded environmental factor.

In the cases discussed above inspection of the diagrams shows clear correlation, but the relationship may not be so obvious if the environmental variation is irregular, and no clear zoning of vegetation is evident. Again we collect two sets of data, one a measure of the vegetation, the other a measure of the environment, but this time from a number of sample quadrats placed symmetrically or at random in the area being studied. In this case a scatter diagram is the easiest way of judging whether there is

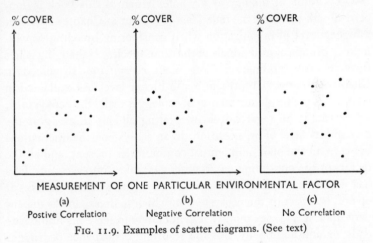

FIG. 11.9. Examples of scatter diagrams. (See text)

correlation. The records for percentage cover, for example, and the level of the environmental factor are plotted for each quadrat, and the interpretation of the diagram is best explained by reference to the hypothetical examples given in Fig. 11.9. Fig. 11.9a shows positive correlation, that is, the percentage cover increases with a rise in the level of the habitat factor. In b we see the reverse, while in c there is no correlation. Theoretically a graph can be drawn to represent the relationship (where there is correlation), the scatter of the points from the line being a measure of the variation due to sampling and other errors. The technique of calculating the line (regression) can be found described in text-books on statistical methods, and might be followed up in schools by teaming up a good mathematician with a biologist to investigate some suitable ecological problem.

12

Production Ecology

Different levels of organization in the living world can serve to define areas of ecological thinking, just as they do in the study of biology as a whole. Much of this book is concerned with plants at the *individual* level, but in Chapter 8 we met a higher level of organization when considering breeding *populations* — groups of individuals of the same species, exchanging genes through cross pollination. The plant *communities* discussed in Chapters 10 and 11 represent a still higher level. As explained in Chapter 1, the fundamental unit of ecology is really the **ecosystem,** which might be regarded as integrating all our thinking at the community level of organization. In an ecosystem, the interactions between all the plant and animal communities present, and all the factors of the non-living environment are taken into account. This has important human implications, for man, whether he likes it or not, is a part of the ecosystem in which he lives, and is dependant upon it. He may modify it deliberately to serve his immediate ends, or unintentionally, but the resulting disturbance of balance can trigger off a chain of unforeseen and far-reaching changes, often to man's detriment. In the past, he could exploit land temporarily, as in the shifting cultivation which has devastated much of Africa, or, where misguided agriculture led to soil exhaustion and wholesale erosion, he could always move on to new territory. But if the world's ever-growing human population is to be sustained, there must everywhere be a wise use of resources for *lasting* productivity: man must learn to manage ecosystems without producing that gross imbalance which leads to disaster in the long run. Ecologists are alive to the urgency of this task, and there is much current research into the functioning of different kinds of ecosystem. This work owes much to the foundations laid by Charles Elton in this country, and by E. P. and H. T. Odum and R. L. Lindemann in the United States. As ecological understanding grows, it is vital that its value in guiding policies of land usage should be widely appreciated. In the words of the American botanist Paul B. Sears (1962), ' — ecological understanding must be built into the wisdom of the race. Man's hope lies not in disregarding his place in nature, but in respecting it.'

GENERAL CHARACTERISTICS OF AN ECOSYSTEM

How do the principles of plant ecology which we have discussed so far fit into the wider concept of the ecosystem? To understand this we must first know more of what goes to make up an ecosystem, its structure, some of its main characteristics and the general way in which it functions. Essentially, any recognizable community of living organisms, together with those non-living substances of their environment which they need, may be regarded as an ecosystem. A beechwood, an area of chalk downland, a saltmarsh, a water meadow, a pond or stream — with their plant and animal communities and the non-living environment with which these interact; all these are distinct ecosystems. Most ecosystems are more or less self-contained (but see p. 229), though of course they are not completely isolated from their neighbours. Obviously a beechwood will have its effects on adjacent downland, and *vice versa*. There is likely to be scrub vegetation, together with its animal communities, along the boundary between the two, and a mingling of both plant and animal species. Some of the animal species, for example a kestrel nesting in a woodland, but hunting chiefly on the downs, may well use, and thus form a part of, both ecosystems. A key characteristic of ecosystems is their marked tendency to stability. The populations of the different species within them influence each other in such a way as to maintain a balance, both in their own numbers, and in their exchange of materials (oxygen, carbon dioxide, minerals etc.) with the non-living part of the system. Thus, an established aquarium may be regarded as a simple ecosystem (and is a very useful model for laboratory studies), but the same cannot be said for a haphazard collection of plants and animals brought back from a pond-dipping excursion and dumped into a tank of water.

Basically, then, an ecosystem is made up of balanced communities of living organisms, together with the substances of the non-living environment which affect their livelihood. Even in a comparatively simple case the living component may include a bewildering number of species from such diverse groups as vertebrates, protozoans, seed plants, bacteria, arthropods, fungi and algae. But as we are mainly concerned with interrelationships of function, we can follow a simple classification of the plants and animals, based on their different roles in the ecosystem. This simple grouping according to **trophic levels,** or general means of nourishment, allows one to see the whole structure in perspective, and to compare widely different kinds of ecosystem.

The first group comprises the **producers** of the ecosystem. These are the **autotrophic** ('self-nourishing') organisms; the green plants which harness light energy and use it to build up their own food from simple inorganic raw materials. All other living organisms, except a few specialized chemosynthetic bacteria, are **heterotrophic,** requiring continued supplies of complex organic compounds. These constitute their food, providing energy to drive them, and a source of building materials, which must include those essential compounds that they cannot synthesize for themselves. Food is of course derived from the excess production by green plants, but some heterotrophs eat plant material directly and others at second-hand. The heterotrophic component of the ecosystem can therefore be subdivided into **primary consumers,** or herbivores in the widest sense, and **secondary consumers,** or carnivores. A further link in the chain is possible: large carnivores preying on smaller ones. These are often called top carnivores, but the term is loosely used also for any large animal at the top of the food chain, even though, like the lion, its diet may consist of herbivorous animals.

This grouping emphasizes a pattern of flow of materials through the ecosystem, from the non-living environment to the producers from these to the herbivore populations, and thence to the carnivores. How is the supply of materials maintained, and where does it all go? Clearly, the chemical weathering of rocks does not proceed fast enough to keep pace with the mineral uptake of green plants, and a system with only producers, herbivores and carnivores could not function for long. Minerals in the soil (or in pond water) would become exhausted, and supplies of essential elements 'locked up' as organic compounds in the bodies of plants and animals. A fourth group, the **decomposers,** plays a vital part in the functioning of the ecosystem by 'unlocking' these bound elements and releasing them in a form which can be utilized again by the green plants. The group is made up chiefly of soil micro-organisms such as bacteria and fungi, but earthworms, beetles, mites and other small arthropods make important contributions. Thus materials from the non-living environment (the abiotic component of the ecosystem) follow a cycle, shown diagrammatically in Fig. 12.1a. After the materials have been taken into circulation by the producers, a fraction is passed on to the primary consumers (herbivores), and from these a fraction goes to the secondary consumers (carnivores). Obviously the herbivores do not consume *all* the producers, neither do the carnivores eat *all* the herbivores, so in due course decomposers will get to work on the dead bodies of all three groups, returning the materials to the non-living environment for the cycle to begin again. Excretory products and faeces from the consumers provide further material for the decomposers to work upon. Whilst most of the elements circulating are stored in the soil between cycles, we must not forget that the pool of carbon (the element that makes up about half the dry weight of most organic matter) is the meagre 0·03 per cent of carbon dioxide in the atmosphere. This carbon dioxide is derived from the respiration of all four trophic groups, and their overall rate of respiration balances photosynthesis to give a more or less constant proportion of carbon dioxide in the air — an example of **homoeostasis** at ecosystem level. This serves to emphasize the importance of the atmosphere as part of a typical ecosystem.

Just as materials are passed on from one trophic level to the next, through the ecosystem, so it is with the energy needed to maintain all the living organisms. There is, however, one crucial difference. Materials *circulate*: they are recovered to be used again and again, but energy can never be recovered; it *flows through* the ecosystem, and is then lost forever. It follows that (leakage apart) an ecosystem can continue to function on a fixed quantity of materials (not a very large quantity if the turnover is rapid), but it will need fresh supplies of energy all the time to drive it. The cycling of materials can be likened to the gearwheels of the mechanism, nothing will happen without a continued input of energy. This is a consequence of the laws of thermodynamics. The first law states that energy can neither be created nor destroyed, though it

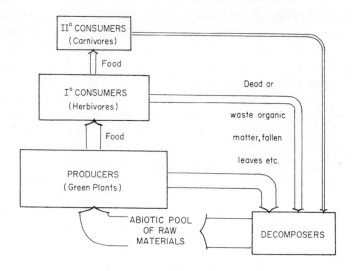

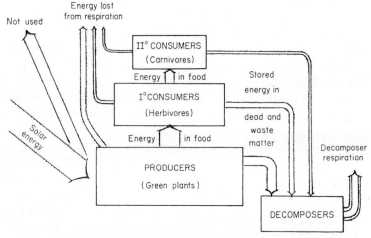

FIG. 12.1. Models depicting (*upper diagram*) circulation of materials in an ecosystem (note key role of decomposers) and (*lower diagram*) flow of energy *through* ecosystem (note: no return from decomposers to producers)

can be transformed from one form to another (e.g. light to stored chemical energy; chemical energy to mechanical energy or heat). It follows that there are fixed 'rates of exchange' between one form of energy and another, as with foreign currencies under stable financial conditions. The value given for the mechanical equivalent of heat is a statement of one such rate of exchange. Each form of energy has its own 'currency' units for measurement, but for ease of comparison it is convenient to use a common unit, whatever form the energy takes. For this the units of heat are employed, either the calorie (1 cal. = heat required to raise the temperature of 1 gm. of water 1 °C) or the kilogram-calorie or Calorie (= 1000 cal.).

The second law of thermodynamics states that no spontaneous energy transformation can take place without some degradation of energy to unorganized or random form. This means, in effect that no energy transformation can be 100 per cent efficient; there will always be a 'loss' as heat *which cannot be used again*, rather like the brokerage charge for changing money from one currency to another. At each level in the ecosystem some energy is 'lost' as useless heat resulting from the numerous energy transformations in respiration and other vital activities. By the time the final decomposer stage is completed, the whole of the energy input has been converted to heat and 'lost'. Fig. 12.1.*b* summarizes this, and it will be noted that there is no arrow from decomposers to producers as in Fig. 12.1.*a*. Nothing is left to feed back into the system.

VARIATIONS IN THE GENERAL PATTERN OF ECOSYSTEMS

The general picture of almost any normal ecosystem would fit into the pattern outlined. The four trophic levels: the uptake of materials from the environment and their circulation through the system to be released again for recirculation; the flow of energy from one trophic level to the next — all these are just as much characteristic of a pond ecosystem as of a woodland, a moorland or a chalk down. Let us look more closely into the pattern and enquire what variation of the general theme is possible.

One might ask whether an ecosystem could function without all four trophic levels being represented. Obviously the decomposers are essential, for even if fresh raw materials could be 'imported' continuously to keep the producers going, there would still be a problem of accumulating waste and dead matter. This scavenging is no straightforward process in which only two or three species are involved. The decomposers form a highly complex community, which is virtually an ecosystem in itself, taking many years to establish. Once aware of this, we can appreciate more clearly the crippling effects of severe soil erosion, which leaves an ecosystem stripped of its whole decomposer community.

Natural communities do exist, in which there is continuous mineral enrichment. Perhaps the most striking case is the marine community along the coast of Peru, where the ocean current brings to the surface a constant supply of mineral-rich water from the depths of the Pacific. On the ocean bed, where this water comes from, the minerals cannot be used, as no light penetrates to these depths to provide energy for photosynthesis. Once the water is brought to the surface where light is available, it is like a richly manured field, yielding enormous crops of phytoplankton, which in turn support vast populations of fish. Fish-eating sea-birds abound, and their droppings, collected from around the nesting sites, furnish the guano which is of world-wide importance as a fertilizer. The food chain has served to concentrate the lost minerals from the bottom of the sea. The guano, in turn, provides imported mineral enrichment, supplementing the nurient pool of the artificial ecosystems of human agriculture. But though in both these examples nutrient minerals are imported into the ecosystem, the decomposer group is still active and of vital importance. The imports of further raw materials simply encourage higher productivity.

Theoretically an ecosystem could function without any consumer organisms. They are indeed sometimes scarce, as in arctic or alpine tundra, but it is hard to think of an example where they are entirely lacking. The excess production by green plants offers many ways of making a living (ecological niches); it seems there is always some consumer fitted to exploit each of these opportunities, whatever specialized adaptation may be needed. Darwin's study of the finches of the Galapagos Islands (Lack, 1953) provides an excellent illustration. E. P. Odum (1963) points out that fire can sometimes act as a regular consumer where there is a tendency to a build-up of dead vegetation, and its effects are not necessarily damaging to the productivity of the ecosystem. It has been shown that in some parts of the United States, periodic fires, started by lightning, are an important ecological factor in determining the climax vegetation; they can almost be regarded as a part of the local climate (E. P. Odum, 1959).

Consequent on their role in the ecosystem, a hazard for consumers, especially secondary consumers, is the tendency for certain toxic substances from the environment to accumulate in their bodies. This poses a growing problem, which demands careful attention. The toxic substances may be poisons deliberately applied by man to control weeds or pests, or simply the waste products of his activities, polluting the environment. In either case, once introduced into the ecosystem, they can have very serious and far-reaching effects, particularly some of the insecticides which are stable compounds and act as cumulative poisons. There is already good evidence of a high mortality of birds resulting from the indiscriminate use of insecticides. This accumulation of toxic substances in particular species is shown dramatically in the case of radioactive isotopes, which can be traced with precision. Radio-ecologists at the American atomic energy plant at Hanford on the Columbia River have made detailed studies of the fate of radio-isotopes released into the river in minute amounts (Hanson and Kornberg 1956; quoted in E. P. Odum 1959). Expressed as a ratio (taking the quantity present in the river water = 1) the concentration of radioactive phosphorus, P^{32}, was found to reach as much as 500,000 in young swallows (feeding off insects from the river), and 200,000 in the yolk of goose and duck eggs; and this despite the short 14-day half-life of P^{32}. Radioactive iodine, I^{131}, reaching vegetation through atmospheric contamination, was found to be concentrated as much as 500 times in the thyroid of jack rabbits, compared with a figure of 1 for the vegetation from which they derived it. The release of apparently harmless concentrations of radioactive wastes into the environment carries obvious risks, but fortunately there seems to be an appreciation of the need for ecological research to guide practical policy in the disposal of these materials. This is applied research indeed, but at the same time, what wonderful opportunities it offers for discovering more about 'community metabolism'. Professor E. P. Odum has himself been concerned in some of this work, and has added an extremely valuable chapter on radiation ecology to the second edition of his textbook (1959).

Rather unexpectedly, the component of ecosystems which is sometimes missing is the producer group. Systems without any green plants (and hence independent of light) can function on a heterotrophic basis, depending on imports of organic compounds for their supply of energy and materials. It seems reasonable to regard them as ecosystems in their own right when the imported organic matter supports a balanced community of consumers and decomposers. The communities found living in the stygian darkness of underground caves provide rather a freakish case, but we have a better example to hand in the decomposer community of, say, a woodland soil. Though part of a wider ecosystem, the soil community is so rich, and the interrelations between species and the environment so close and complex, that the whole can be regarded as a heterotrophic ecosystem in itself. One has simply to examine a handful of leaf mould, spread out under a bench lamp, to realize how much more is present than the fungi and bacteria traditionally associated with decomposition. Besides earthworms and the larger arthropods such as centipedes, millipedes, beetles and their larvae, one can find a variety of fly maggots, spiders, springtails, mites and false scorpions. Extraction techniques, even with crudely made equipment, show the populations to be very large indeed (Kevan, 1962). Densities of 40,000 arthropods and 2,000,000 nematodes per square metre, all within the top few inches of soil, are quite normal. The detailed relationships between these creatures are not fully understood, but the general outline seems clear enough. Different species of fungi and bacteria are the key organisms, as they alone can make all the enzymes needed to attack the more resistant substances, but detritus feeders help to speed the process by reducing the litter to smaller fragments with relatively larger surface. In feeding on the litter, the detritus-feeders must derive some nourishment from it (perhaps with the help of their intestinal microflora), but some are thought of rather as fungus feeders. Detritus-eating mites are so abundant that their droppings may form an appreciable bulk of the litter, providing food for other species of mites, springtails and nematodes. This array of small animals provides an opportunity which carnivorous species do not miss. There are predacious mites, which eat their detritus-feeding fellows, carnivorous centipedes, beetles, spiders and false scorpions. Even the fungi join in: Duddington's work (1957) has shown that in a number of species the hyphae form gin traps which close on unwary nematodes, holding them fast while their body juices are absorbed.

What at first sight seemed a simple decomposition process turns out to be a complex involving a bewildering variety of different organisms. This richness is made possible by the abundant stored energy that is fed into the system every autumn in the falling leaves, and also by the gradual pace of their decay. Taking two or three years to complete, this allows a steady flow of energy at all seasons, and under these relatively stable conditions a balanced ecosystem can develop. Something approaching this state of affairs may also arise with a long-term decomposition 'project', such as a decaying tree stump. There is, however, the important difference that the system is working on capital, which will all be used up, rather than on a regular income of dead leaves. Short-lived decompositions, as in animals carcasses or cowpats show such rapid changes in consumer populations that they can hardly be regarded as established ecosystems.

Having examined ecosystems in general, we can now turn to the particular. What is it that distinguishes one ecosystem from others? How can we analyse it? What information do we need to describe it? What features must we try to measure if we are to define it accurately? In seeking answers to these questions, three kinds of data are important:

Data relating to (i) the species structure, (ii) the structure in terms of the different trophic groups and the reserve pools of inorganic materials, and (iii) the functioning of the ecosystem, especially in terms of flow of materials and energy.

SPECIES STRUCTURE IN ECOSYSTEMS

Broadly speaking, the **species structure** of an ecosystem can be taken to mean the number of species present at each trophic level, and their relative abundance. Further details about the ecological niches occupied by different species, will allow a finer analysis. All this information may be collected as a preliminary step in the study of trophic structure, but its direct value lies in what it can tell us about diversity, successional changes and stability of the ecosystem. For example, a demanding habitat such as a rock face or a mobile sand dune can support only those few plant species whose range of tolerance allows them to survive there. In such a limited plant community there are few ecological niches available for primary and secondary consumers, and the ecosystem which develops will be poor in species. If any misfortune, such as a bad season or an introduced parasite, strikes the population of one species, there will be few others to 'buffer' the effect, and the balance of the whole ecosystem may be gravely upset. A species-poor ecosystem tends to lack stability. But usually the plants bring about a gradual amelioration of the habitat (through soil formation, shading etc.), and the pioneers are replaced by other species. As the environment becomes progressively less demanding, not only do we find different species of plant, but a much greater diversity, and these changes are reflected in the animal populations which the vegetation supports. Both at primary and secondary consumer levels the pioneers are replaced by a wider variety of species occupying the new ecological niches as they become available. Thus the seral changes in plant communities discussed in Chapter 10 bring about a wider succession in which the whole ecosystem becomes more complex, more stable, and up to a point more productive. The increased stability is a direct result of richer species structure, with its many alternative paths in the food web. Any sudden increase or decrease in the population of one species can be compensated by changes in others occupying similar ecological niches. This homoeostatic mechanism, or buffering effect is an important attribute of well established ecosystems, making them resilient to changes, yet still adaptable. The increase in productivity arises from fuller utilization of resources by the plants, allowing more effective capture of the light energy falling on the area. But in climax communities, although there is maximum bulk of organic material as standing crop, it seems that the stage of maximum productivity is often past. The high proportion of energy 'locked up' in unproductive wood must contribute to this. Also, it is believed, forest trees may produce more layers of leaves than the available light justifies, the innermost ones being so heavily shaded that they cannot balance their respiration.

Even in a mature, species-rich ecosystem the bulk of the community is made up of relatively few common species. The diversity which succession brings depends on those that remain uncommon. Their importance lies in the reserve of adaptability which they represent: rather like the gene pool of an adaptable population. This is seen at each trophic level, though among the producers there may be a fall in numbers of species as the succession proceeds. This is the outcome of dominant species (e.g. vigorous

grasses or trees casting heavy shade) imposing unfavourable conditions on other plants in the community. As an instance, the
disappearance of rabbits (which held up and diverted plant succession) has led to a fall in species diversity in many areas of chalk
grassland now dominated by vigorous meadow grasses such as
cocksfoot and false oat. The growing species diversity which usually
accompanies succession may not be evident from records of particular taxonomic groups. For example, both grasses and grasshoppers
are better adapted to the conditions of open habitats, and one would
expect to find far fewer species in woodland communities.

One of the earliest signs of damage to an ecosystem from pollution of the environment is a decrease in the diversity of species.
The least tolerant species die out first, followed by others as the
pollution builds up. The richness of the lichen flora on tree trunks
has been used by the Dutch botanist Dr. Barkman as a sensitive
index of atmospheric pollution in industrial regions. By noting
which species are absent he can assess the severity of the pollution.
There is commonly a 'lichen desert' extending for miles downwind
from the source of pollution, with only the most resistant species,
such as *Lecanora conizaeoides*, penetrating nearer. Many species
have become extinct in industrial regions of Western Europe
during the last hundred years. Rivers or lakes exposed to pollution,
and areas of land threatened by indiscriminate use of insecticides
or weedkillers are other cases where we need early warning of
damage to the ecosystems. Various **diversity indices** have been
devised for detecting small changes in species structure (or for
comparing species structure in different ecosystems). A simple
index which has been widely used is the ratio:

$$\frac{\text{Total number of species recorded}}{\text{Log. number of individuals counted}}$$

To determine this, counts of individual organisms are made from
a number of sample areas, and a record is also kept of the running
total of different species included. If one plots a graph of the
running total of species against the log. of total number of individuals, the result is roughly a straight line. This means that the
slope, which is called the diversity index, is broadly speaking in-
dependant of sample size. Determinations may be based on organisms within a single trophic group (though producers are a less

reliable guide), or on any wide taxonomic group, such as arthro-pods or insects, or indeed on all groups together. It is the changes in the index with time (or differences, if one is compar-ing ecosystems) which are important, rather than any absolute values.

TROPHIC STRUCTURE OF ECOSYSTEMS

The second variable by which different ecosystems may be dis-tinguished is their **trophic structure,** or, crudely speaking, the relative sizes of the 'boxes' in Fig. 12.1. This could be based on numbers of individuals per unit area at each trophic level. As an example Fig. 12.2.*a* illustrates numbers of organisms in a grass field community. The producers include grass and other herbs; the herbivores are all invertebrates; carnivores include spiders, ants, predatory beetles etc; and the top carnivores birds and moles. No estimate is given for the decomposer population, but the figures would certainly be very large indeed. The diagram presents the data as the familiar pyramid of numbers, which adds point to the Chinese proverb 'One tiger: one hill'. While expressing a general truth, the pyramid of numbers is of no help in making comparisons between ecosystems, because it takes no account of the size of organisms. If the producers were beech trees the 'pyramid' would sit on a point, but if they were diatoms it would be very broad-based indeed. A better measure for comparison is the weight of living matter, or **biomass,** at each trophic level. Estimated at some particular time this would be the **standing crop** at that time, a term applied both to plants and animals at any level in the ecosystem. Biomass can be expressed as live-weight or (better) dry-weight, or in various more sophisticated ways such as total protein or organic nitrogen. Returning to Fig. 12.1.*a* we will take the 'boxes' to represent relative biomass at different levels. Though not drawn to scale, the diagram clearly indicates that the standing crop of herbivores has a smaller biomass than that of the plants, and the standing crop of carnivores is smaller still. We have a pyramid of biomass. Studying the diagram, we might say that it *must* take the form of a pyramid: the carnivores cannot eat up all the herbivores, neither can the herbivores consume the whole crop of plants, so it *seems* that there must be greater biomass at the con-sumer end. This argument is unsound, however, for it takes no

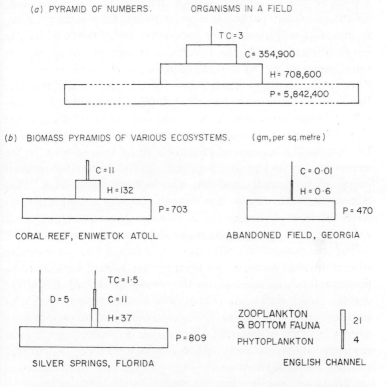

FIG. 12.2. (*a*) Numbers of organisms in a grass field arranged according to trophic levels — a pyramid of numbers. (Decomposers not included.) (*b*) Examples of biomass pyramids based on widely differing aquatic and terrestrial ecosystems. Drawn to approximately the same scale; figures represent biomass dry weight in gm. per square metre. P = producers, H = herbivores, C = carnivores, T C = top carnivores, D = decomposers

From Odum, 1959; courtesy of W. P. Saunders Co., Philadelphia

account of the *rate* of flow of materials from one trophic level to another. 'Pyramids of biomass' are indeed normally pyramid shaped, but where productivity is very rapid, as in marine plankton, an inverted pyramid is possible, with a smaller biomass of producers supporting a larger biomass of primary consumers. Fig. 12.2.*b* shows examples drawn from different ecosystems, and it will be seen from the differences that biomass can be useful as a measure in comparative studies. In pioneer communities one expects to find a relatively smaller standing crop of consumers, and this is evident in the narrower bands at the top of the pyramid, as compared with the broader apex in the case of the coral reef, an old established community. Inverted pyramids are typical of open water communities, where the producer group is made up entirely of phytoplankton. It is characteristic of very small organisms that they have a much higher rate of metabolism than a corresponding bulk of larger organisms. As a result, a relatively small standing crop of phytoplankton can increase so rapidly that it will support the grazing of a biomass of consumers larger than its own. In the same way, the decomposer population in the soil tends to consist largely of very small organisms with high metabolic rates. What data are available suggest that, though numbers are large, decomposer biomass is unexpectedly small compared with what it achieves (note biomass pyramid for Silver springs in Fig. 12.2.b).

Plankton populations, with their rapid rate of turnover, are also subject to wide fluctuations affecting the balance between producers and consumers. Studies of a pond at Brentford, and of the disused canal at Tiverton have both shown an approximately monthly cycle, with peaks of phytoplankton and of zooplankton alternating. The evidence of direct counts on limited samples was supported by an indirect method in which both dry-weight and relative chlorophyll content were determined from standard samples taken weekly. The chlorophyll estimation was based on light absorption of a standard acetone extract. The dry-weight figures, which included both plant and animal plankton, provided an index of total respiration, while the density of the chlorophyll extract was taken as a measure of photosynthetic capacity. Plotted against time, the two curves fluctuated in opposition, peaks of maximum dry-weight coinciding with minimum chlorophyll content, whilst the dry-weight fell off during the ensuing build-up of chlorophyll. Though many other factors undoubtedly play their part, it is tempting to see in these results the oscillations produced by a rather crude feedback mechanism — the zooplankton building up until there is overgrazing, which reduces the phytoplankton to such a low level that the consumer population crashes, giving the phytoplankton a new lease of life. Equilibrium cannot be reached in the short span of one growing season, and in the following spring the performance starts all over again.

ECOSYSTEM DYNAMICS — FLOW AS BIOMASS

As agricultural crops are usually accumulated food stores harvested when mature, we tend to think of 'the crop' just as the *amount* taken at harvest. It is easily overlooked that, used in this sense, the word 'crop' expresses a *rate* of production, for the harvest represents what was accumulated in a known period of time (i.e. production during the growing season, less what was eaten by the pests). But the ecological term 'standing crop' means simply what stands there at any particular time, taking no account of how long it took to accumulate, nor how fast it is being eaten up. The standing crop of grass in a field may be small because it is an unproductive pasture, or because it has been heavily grazed, or because the estimate was made before the growing season. Herein lie the shortcomings of standing crop biomass data. They provide useful basic information about structure, but tell us little of the working of ecosystems. We could as well try to reconstruct the steps and tempo in a Scottish dance from a single snapshot.

What kind of data do we then need? In a stable ecosystem we might expect to find the standing crop at each level fairly steady, with primary production balanced by grazing, and the increase in herbivore tissue offset by the harvest which the carnivores are taking. The activities of decomposers could serve as a kind of compensating device through their effect on the supply of inorganic materials to the producers. But this balance could be achieved either with sluggish metabolism at all levels, or with the organisms living at a furious pace, their growth, reproduction and rate of turnover of materials all at a maximum. It is, then, the dynamics of ecosystems (the 'arrows' in Fig. 12.1) that we should investigate, and the first step is to obtain data on **productivity,** or *rate* of production, at each level. To speak of the productivity of a consumer may seem a contradiction in terms, but the concept is really quite straightforward. We may think of it as the increase in herbivore tissue which is 'offered' to the carnivores for consumption over a given period of time, or the rate of supply of carnivore tissue to the decomposers.

There is no close relationship between standing crop biomass and productivity, for the latter is influenced by the size of individual plants and animals. As we have already noted from plankton communities, small organisms generally live at a higher metabolic rate than larger ones, hence the inverted pyramid of biomass. Kleiber (1961) quotes figures which illustrate this nicely. Starting with a ton of hay, we could feed it either to a bullock or to 300 rabbits (having about the same aggregate weight). The bullock, he says, would get through the hay in 120 days, giving an increase of about 240 lb. of meat, but the rabbits would consume it in 30 days, yielding the same increase of 240 lb. of meat, though in one quarter of the time. The productivity of the rabbits is thus four times that of the bullock.

The simplest measure of the productivity of an ecosystem is in terms of the rate at which animals at the top of the food chain can be harvested. Managed fish ponds provide an example, and one can assess the effect of adding fertilizers to the water from the productivity of fish. In the same way, the statement that a certain area of grassland will 'carry 0·8 head of cattle per acre' is really an indirect expression of the productivity of the pasture (in terms of the continued harvest of meat or milk one would expect from 0·8 cattle). But unless there are records showing sustained productivity over a long period, this is scanty information on which to base management policy. How are we to know that the ecosystem is not being overgrazed or overfished, leading to permanent damage to its structure if its homoeostatic mechanism is overstrained? Details of productivity at other levels are needed too.

In pond ecosystems there is a simple method by which productivity of the combined phyto- and zooplankton can be estimated. This is the 'light and dark bottle method' described in E. P. Odum (1963). If we assume that basic carbohydrate metabolism predominates, we can summarize the production and respiration processes of the plankton in terms of the equation:

$$2H_2O + CO_2 + (114\ Cal) \underset{\text{Respiration}}{\overset{\text{Photosynthesis}}{\rightleftharpoons}} \underset{MW=30}{(CH_2O)} + \underset{MW=32}{O_2} + H_2O$$

As photosynthetic activity is directly related to oxygen release, and respiration to oxygen consumption, changes in oxygen concentration within a closed system provide a means of assessing the cumulative balance of these two processes in the plankton community, i.e. its productivity. Essentially, we set out in this method to measure the changes in oxygen concentration over

24 hours in two water-plus-plankton samples, sealed in glass bottles, and suspended in the pond at the depth at which the samples were collected. In other words, we are simply isolating two samples of pond from the surrounding water (to prevent any oxygen exchange), but otherwise keeping them under the natural conditions of the plankton community — with the exception that one of the bottles is wrapped in aluminium foil, so that its contents remain in complete darkness. In this bottle (the 'dark bottle') only respiration of the plankton can take place; in the 'light bottle' both respiration, and photosynthesis during daylight hours. A third sample must be taken at the time of planting the bottles; from this the initial content of oxygen dissolved in the pond water is determined. The other two bottles are harvested after 24 hours, and the oxygen content of the water in each is estimated. The temperature of the water should be recorded at the time of taking out each sample. For oxygen estimations one can use the Winkler technique (see Appendix, p. 271). Using data obtained from further sets of bottles suspended at different depths in the pond, one can build up a picture of plankton productivity in a column of water beneath 1 square metre of surface, and so arrive at a figure for the whole pond. The treatment of data is best explained by reference to actual results obtained at Slapton Ley in July.

TABLE 12.1. *Oxygen concentrations recorded using light and dark bottle method, Slapton Ley, Devon, July 1966*

Samples taken at ½ metre depth

		Light bottle	Dark bottle
Initial oxygen content	(mg./litre)	10·13	10·13
Content after 24 hours	(,, ,,)	11·10	9·64
Change in oxygen content	(,, ,,)	+0·97	−0·49

The loss of 0·49 milligramme per litre of oxygen in the dark bottle represents respiration of the plankton over 24 hours. The light bottle shows a net gain of 0·97 mg./litre of oxygen from photosynthesis in excess of respiration. Referring back to the equation on p. 238 it is clear that the oxygen figures can be converted to weights of carbohydrate gained or lost by multiplying by 30/32 (the ratio of the molecular weights of carbohydrate and oxygen). Further, concentrations expressed in mg./litre are numerically the same as grams per cubic metre, so we can say that the net photo-synthesis, or net production of the plankton in one cubic metre of water in 24 hours was $0.97 \times (30/32) = 0.91$ gm. carbohydrate. The dark bottle results show that a similar amount of plankton used $0.49 \times (30/32) = 0.46$ gm. carbohydrate in 24 hours for respiration, so there must have been a gross production of $0.91 + 0.46 = 1.37$ gm. in 24 hours from photosynthesis of the phytoplankton. Measurements from other sets of bottles suspended at greater depths, show a decrease in productivity, which one would expect, because of the fall in light intensity. By estimating the mean productivity from a column of water with surface 1 square metre one can arrive at a figure for the lake as a whole. In the case of Slapton Ley the net productivity on this occasion was of the order of 7000 kg. of carbohydrate daily; enough to support quite a population of fish.

It is interesting that rates of primary production (i.e. net photo-synthesis) from open water, well stocked with nutrients, are of the same order as from fertile land. Evidently the medium, whether seawater, freshwater or land, does not greatly influence the performance of green plants adapted to grow there. Their efficiency depends rather upon the extent of photosynthetic sur-face which they can expose the intercept light. This may be limited by the availability of raw materials needed to build the plant body. The tiny structures of unicellular algae, which are thought to account for some 90 per cent of the world's photosynthesis, achieve a maximum of surface for a minimum of material, yet in the oceans and in many lakes it is shortage of minerals that restricts productivity. The suggestion has been made that nuclear power might be employed to pump up mineral-rich water from the ocean depths, simulating the natural upwelling along the coast of Peru (p. 228). On land it is more often low rainfall and low temperatures which limit the availability of minerals, and the effective growth of plants. Most desert soils contain rich supplies of nutrients, but, lacking water, the vegetation can make but little use of them in building leaf area to intercept the abundant solar energy.

The effect of increased leaf area on productivity seems so obvious that we may easily fail to realize its extent. The results of a simple study of individual sunflower plants are revealing. Determinations of daily net photosynthesis were made by comparing the dry-

weight of discs cut from leaves before and after six hours in sun-
shine, or in darkness. These were then related to the 'building
capacity' of a sunflower plant (in terms of new leaf area) throughout its early life, and compared with its actual performance. The
results are admittedly a first approximation, as the calculations
assume that all production went into leaf building, and no allowance has been made for the part played by the roots. They are
summarized in Table 12.2.

TABLE 12.2. *Data on productivity and growth in leaf area of sun-
flowers*

Stage of development of plant (No. of leaves developed)	Area of last full-grown leaf (cm.²)	Total leaf area of plant (cm.²)	Production capacity (mg./day dry matter)	Equivalent production as leaf area (cm.²)
2	4·7	9·3	7	2
4	13·5	34	26	9
6	25	84	64	21
10	99·5	395	300	100
15	175	1212	920	307

Gross photosynthesis	1·62 mg. dry matter per square centimetre leaf area daily				
Respiration	0·86 ,,	,,	,,	,,	,,
Net photosynthesis	0·76 ,,	,,	,,	,,	,,

It will be noticed that there was a regular increase in the area of
successive leaves, but by leaf No. 12 a constant maximum size of
about 175 cm.² was reached. Disregarding any food store remaining from the seed, we see that the plant's total production capacity
at the 2-leaf stage could build only 2 cm.² of leaf per day. At this
rate it would take more than 12 weeks to build a 'full-sized' leaf
of 175 cm.² At the 4- and 6-leaf stages the production capacity
had greatly increased, but it would still require about ten days to
make a full-sized leaf. Instead, smaller leaves were produced, their
size nicely adjusted to account for 2 to 3 days' production at the
stage at which each was formed. By the time the 11-leaf stage was
reached, the plant's total productive capacity was sufficient to
build a full-sized, 175 cm.² leaf in less than two days. It is interesting that the observed time intervals between completion of successive leaves agree well with these calculated times. Two things
stand out from this simple investigation. First, the enormous

increase in production capacity (more than one-hundredfold) within a few weeks that is made possible from added leaf area. Secondly, the profound effect on total productivity that would result from any check on the *rate* at which the plant can make new leaf tissue (e.g. shortage of soil minerals, or a daily period of wilting). What is true of individual plants applies also to whole plant communities, and helps us to understand the variations in productivity of vegetation in different parts of the world (Fig. 12.3).

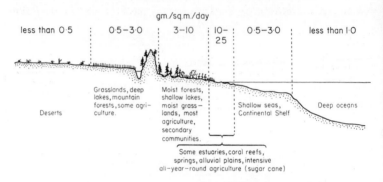

FIG. 12.3. World distribution of primary production, in gm. dry matter per square metre per day (based on average daily rates of gross production in major ecosystems)

Redrawn from Odum, 1959; courtesy of W. P. Saunders Co., Philadelphia

From *rates* of primary production (N.B. *not* standing crop) one can make rough predictions of productivity at other levels in the ecosystem, for it is generally the case that productivity at any level is *approximately* one tenth of that shown by the preceding trophic level. Suppose, for example, that a given area of land or water produces 1000 kg. of plant dry-matter each year, then one could expect an annual production of around 100 kg. from the herbivores, and 10 kg. from the carnivores which the area supports (all figures dry-weights). If there is a further top carnivore link in the chain, its productivity would be only of the order of 1 kg. per year. It is in the step from plants to herbivores that this factor of $\frac{1}{10}$ varies most, because in some ecosystems (e.g. forests) a very large proportion of the primary production is not used by the herbivores, but goes direct to the decomposers.

To determine these figures, studies of the dynamics of terrestrial ecosystems must generally begin with periodic estimates of standing crop biomass at the different trophic levels. Data on respiration rates, food consumption, growth and reproduction rates etc. must then be collected, in part from individual studies, in order to assess how food is used and the rate at which it passes from one trophic level to the next. Owing to the great difficulty in obtaining anything like complete data on all the consumer populations, it is often expedient to concentrate on a particular food chain within the ecosystem. The way in which the general picture can be pieced together is well illustrated by Golley's (1960) studies of the dynamics of a food chain in an abandoned field community in America. Data were collected over a year (May 1956–May 1957) on a food chain made up of the field vegetation (producers), meadow mice (herbivores) and least weasel (carnivores). Below, greatly simplified, is a summary of the work involved at each level:

Producers (Grass and other herbs)

Net rate of increase in biomass (after grazing) through the growing season, estimated from periodic harvesting of sample plots. Shoots were harvested by clipping to ground level, roots by washing out from known volume of soil. Living material was sorted out from dead, and dry-weights determined separately.

Respiration rate of vegetation in the field, determined by measurement of oxygen/carbon dioxide exchange in small sample plots covered by oil drums.

Consumers–herbivores (Meadow mice)

Population density estimated at intervals by live-trapping, and mark-and-release methods.

Age distribution and production of young estimated from data on size, weight and breeding condition of trapped individuals.

Growth rate at different age categories estimated from weights of individuals trapped repeatedly.

Respiration rate of mouse population estimated from laboratory experiments with individuals.

Food consumption and assimilation estimated from laboratory feeding trials, analyses of stomach content of trapped individuals, and determination of weight of faeces produced.

Consumers–carnivores

Similar data to that collected for the mouse population.

From these data we can build up a flow diagram of biomass, in which the boxes represent standing crop, and the width of the 'pipes' the rate of flow (though not drawn to scale). Rates of flow, whether production of new tissue, consumption or respiration usage, are expressed in terms of biomass per year, as daily values will differ so widely according to season (Fig. 12.4). The rate of

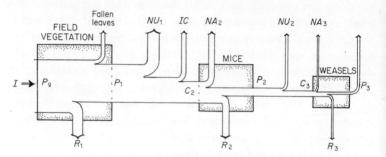

FIG. 12.4. Flow diagram of biomass along a food chain (based on Golley, 1960). For explanation see text

increase in vegetation biomass, estimated from harvesting the sample plots, represents only that part of the production which was not used (NU_1 in Fig. 12.4). To obtain net productivity (P_1) we must work backwards in the diagram, adding the biomass consumed annually by the mice (C_2) and also the consumption by smaller invertebrates (IC) which was not assessed; though the error is to some extent compensated by the presence of these invertebrates in the mouse diet. These three components together give P_1, the net productivity of the vegetation, which is available for food. Actually the mice eat only the green tops, which made up about 20–25 per cent of P_1, and they took less than 5 per cent of these. By adding to P_1 the losses due to respiration (R_1) and making allowance for dead and fallen leaves during the year, we can arrive at gross productivity for the vegetation (P_g), its total photosynthetic achievement over the year.

Turning to the mouse population, the annual consumption of plant matter (C_2) does not all go to make mouse tissue. About one third of the food is never assimilated, but returned to the environment as faeces (NA_2), and a considerable proportion is used in respiration (R_2). The remainder appears as the productivity (P_2) of the mouse population, in the form of heavier individuals and increased numbers.

The mouse productivity (P_2) is available for the weasels, possibly a little augmented or reduced by migratory movements of mice in or out of the experimental area. But the weasels' food consumption (C_3) leaves a considerable population of 'unused' mice, which can eventually 'die in their beds' (NU_2). Before doing so, however, they have a vital role to play in maintaining next year's production. As with the mice, some of the weasels' food consumption is not assimilated, though the proportion here is smaller (NA_3), and some is used in respiration (R_3). In the absence of larger carnivores, the whole weasel productivity (P_3) would be classed as 'unused'. We must not forget, however, that everything in the 'unused' and 'not assimilated' categories goes to maintain the decomposer component of the ecosystem, and the materials are released again, ready for intake (I) by the producers.

This study helps us to appreciate how feedback mechanisms can operate to keep the various populations steady. There is plenty of food for a larger mouse population, but predation by weasels keeps their numbers down. Apparently the weasels have no important predators, but by the time the pipeline reaches their 'box' it is quite narrow, indicating that food supplies are limited. Any sudden increase in weasels would push up their food consumption (C_3), and so reduce the supply of unused mice (NU_2) upon which maintained productivity of the mouse population depends. The long-term effect would be to reduce P_2, making the pipeline narrower still, and thus checking the weasel population. This kind of buffering or homoeostasis is well developed in a complex ecosystem, with its many varied populations; such species-rich communities show great resilience to change. It is in the comparatively simple ecosystems of the arctic tundra, or in plankton communities, or the artificial systems of human agriculture that one finds population explosions. The periodic upsurges of lemming populations and the fluctuations in numbers of arctic hares are well-known examples of instability for which man is not responsible. Monoculture of crops invites disastrous plagues of pests and fungal parasites, such as Colorado beetle, the coffee rust in Ceylon and potato blight in Ireland. Routine control of these organisms is a chore in agriculture adding to production costs, yet within the varied contexts of their natural habitats they are held in check by the buffering of the ecosystem. Species introduced by man into other continents either deliberately or by accident, may sometimes erupt with far-reaching consequences especially where natural ecosystems have been modified (Elton, 1958).

ECOSYSTEM DYNAMICS — ENERGY FLOW

Much can be learned about the dynamics of ecosystems from studies of the flow of materials in terms of biomass. But are we justified in equating biomass of mouse tissue with biomass of plants or weasel droppings? If these are not strictly comparable (which is the case) how close is our approximation? The difficulty can be overcome by converting biomass to energy units (Calories), a simple process of multiplying biomass (gm.) by the calorific value of the substance concerned. Calorific values are the exact rates of exchange for converting to a common currency of energy units. Approximate calorific values can be determined using the apparatus and technique described in Year III of the Nuffield Foundation Project's O-Level Biology Text (1966). Some of Golley's (1960) figures,* quoted below, indicate the order of approximation that use of raw biomass data involves.
Mean calorific values:

> Green tops of herbs and grass: 4·08 Cal/g. dry-weight
> Roots of   „   „   „ : 3·30 „   „   „   „
> Mouse tissue   ⌠: 4·65 „   „   „   „
>   ⌡or 1·37 Cal/g. live weight

Far more important than the role of energy as a universal currency is the fact that it is fundamental to the functioning of ecosystems. It drives the circulation of materials, and its rate of flow through the system is an index of community metabolism. Productivity of the ecosystem as a whole depends upon the percentage of incoming solar energy that the plants can harness in photosynthesis, and the efficiency of the energy transformations between different trophic levels. An energy flow diagram will therefore be our most reliable guide to ecosystem dynamics. For this Fig. 12.4 would serve, if the biomass values were converted to Calories. It has been developed in Fig. 12.5 to give a more complete picture.

In following the course of the energy through the system, we

---

* Golley's final results are, of course, expressed in terms of energy flow; biomass values have been given in earlier discussion simply to help clarify a step in the explanation.

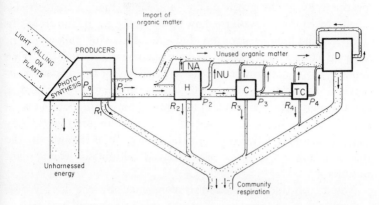

FIG. 12.5. Generalized energy flow diagram for an ecosystem (adapted from Odum, 1957)

are interested particularly in the efficiency of energy transformations, which show not only in the narrowing of the pipelines representing energy flow, but also in the losses through respiration at each stage. All the incoming energy must be accounted for at the 'respiration sink', unless there is any export (which, to avoid complication, has not been shown in the diagram). Dead leaves washed down a stream, or blown away from the study area would constitute export; so would emigration of mammals as part of their territorial behaviour regulating population density. In the same way there can be imports into an ecosystem. Dead leaves may supply much of the energy income of a stream community, passed on to other levels through the detritus feeders. Bread fed to ducks in the local park pond can play a similar role.

Returning to Fig. 12.5 we must look first at the efficiency of the plants as energy converters. The annual income of radiant energy received by a square metre of ground can be measured, using some kind of radiometer. The diagram shows that only a part of this energy income is harnessed, but it cannot conveniently show how small a part this is. Even when allowance is made for the UV and infrared energy which play no part in photosynthesis, we find that less than 2 per cent of the available energy is converted into bound chemical energy — the plants' gross productivity (P_g). Of the rest, some is reflected, some passes through the leaves and much goes to evaporate water, promoting the transpiration stream. The gross productivity of the green plants represents, then, a mere trickle of energy compared with the flow reaching the earth, and from this a further fraction is diverted down the heat sink following the internal conversions which we call plant respiration (R_1). What remains is the net productivity (P_1), available for the primary consumers. This may seem an inefficient performance by the plants, but they nevertheless maintain the entire animal kingdom on what they have produced in excess of their own requirements. Any imports of energy flowing into the system at this point in the diagram are, of course, also derived ultimately from plant production.

Not all of this production is consumed by the herbivores; indeed, in forest ecosystems they take only a very small proportion, and the bulk goes to the decomposers as leaf litter and rotting wood. There is thus, a dividing of the ways here, and energy flowing through the ecosystem may follow either of two main channels. First the **grazing channel,** taking that proportion of the primary production which is eaten by the herbivores, and following the normal course of the food chains. At each trophic step about 90 per cent of the flow is diverted to respiration and to the decomposers, leaving only a small trickle for the top carnivores. It is for this reason that one rarely finds more than four links in a food chain. The second channel, known as the **detritus channel,** starts with the fall of those leaves left by the consumers, and receives tributaries of unused energy from each consumer group. Eventually the entire flow is spent in the heat losses from decomposer respiration in the soil (or the bottom sediments in aquatic ecosystems). It is instructive to estimate the contribution made by a single tree to this channel in its annual crop of leaf litter. From leaf counts, using rough sampling techniques, and determinations of calorific value, it is not difficult to arrive at an approximate figure. One can even make a crude assessment of the flow to the grazing channel by estimating the percentage area of leaves eaten away.

This pattern, with its two channels of energy flow, is common to all ecosystems. Forests are exceptional in the high proportion of the available energy that follows the detritus channel. By contrast, the main energy flow in systems based on phytoplankton usually follows the grazing channel. These two types may be compared in Fig. 12.6.

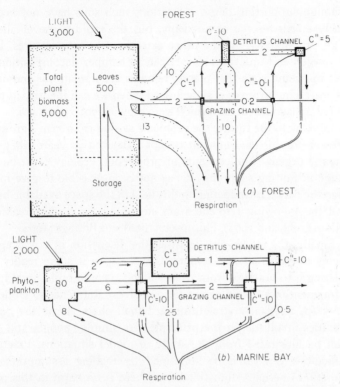

FIG. 12.6. Energy flow diagrams for a forest and a marine ecosystem, illustrating grazing and detritus channels of flow. Drawn roughly to the same scale; standing crop biomass as average kilocalories per square metre for an annual cycle, and energy flow in kilocalories per square metre per day

Adapted from Odum, 1963; courtesy of Holt, Rinehart and Winston, New York

In previous diagrams the organisms of the detritus channel have been represented as a single terminal box, labelled 'decomposers', though it was explained earlier that the soil decomposer community could be thought of as an ecosystem depending on imported organic matter. This is brought out in Fig. 12.6, where the detritus channel is shown flowing through a box representing detritus feeders to a further trophic level of consumers preying upon them. The two together represent the decomposer component. The pace at which the decomposers work varies greatly. Under the ideal conditions of a tropical rain, forest decomposition is completed so rapidly that the soil contains little leaf mould.

Though supporting dense vegetation, such soils have not necessarily a rich reserve of nutrients, but the rapid turnover allows high productivity. At the other extreme, low temperatures, lack of oxygen and unfavourable pH can so hamper decomposition in wet moorland soils that plants' remains accumulate for centuries as peat. The turnover is slow indeed when we can identify in peat today Sphagnum mosses which were growing 7000 years ago.

Looking to the future, the problems arising from man's growing pressure on his environment cannot wait even 70 years; they are already urgent. The concepts of production ecology must be his guide in finding solutions. It has not been possible to give more than the barest introduction to this field in the space available here, and the following references are suggested for further reading: Odum (1959 and 1963), Phillipson (1966) and Billings (1964).

Finally, the reader may feel that the discussion in this chapter seems too abstract and remote. Can we give it some reality by relating it to practical work from which we could build up at least a fragment of a flow diagram for ourselves? Reference to Golley's work (p. 243) on a straightforward food chain shows the complexities involved, yet the principles are simple enough, and can well be illustrated from the study of model situations. Practical difficulties can often be simplified by making assumptions — sometimes perhaps unjustified, but there is no harm in this provided we realize what we are doing, and use our results to illustrate broad principles rather than drawing unwarrented conclusions from them. One such simplification for purposes of calculation is to work all conversion values on the basis of the carbohydrate equation given on p. 238. Making this assumption, we can use measurements of oxygen, carbon dioxide or carbohydrate (dry-weight) for determinations of productivity, whichever is best applicable. (30 gm. of carbohydrate corresponds to 32 gm. (or 22·4 litres) of oxygen; or to 44 gm. (or 22·4 litres) of carbon dioxide; or to $686/6 =$ approx. 114 Kilo-calories.) If we wish for further refinement, respiratory quotients can be checked, but experimental errors in other aspects of the study may well be so great as to make this pointless.

Let us first see how the determinations of plankton productivity by the light and dark bottle method (p. 239) fit in with the flow diagrams in Figs. 12.4 and 12.5. We are here dealing with both phyto- and zooplankton together, so the respiration for 24 hours, derived from the dark bottle results, represents $R_1 + R_2$. We can say, then, that $(R_1 + R_2) = 0.46$ gm. carbohydrate per cubic metre of water per 24 hours, or, taking the calorific value of the carbohydrate as 38 Kilo-calories per gm., $(R_1 + R_2)$ accounts for about 17 Kilo-calories per cubic metre per 24 hours. Similarly, the net productivity of the plankton as a whole (that which is available as food for larger organisms) was 0.91 gm. carbohydrate per cubic metre of water per 24 hours, or $0.91 \times 38 = 35$ Kilo-calories of energy fixed per cubic metre of pond water daily. In the flow diagrams this is made up of two components: P_2, the productivity of the herbivores (here herbivorous zooplankton) and NU_1, that fraction of the phytoplankton not eaten by the zooplankton. These we cannot easily separate. The gross productivity, arrived at by adding respiration and net productivity, must obviously be the work of the phytoplankton alone, and appears as P_g in the flow diagrams. Expressed as energy fixed in 24 hours, this is $1.37 \times 38 = 52$ Kilo-calories per cubic metre of water.

Thus, from the results of the light and dark bottle experiment we can begin to build up something of the general picture. Separation of the plant and animal plankton — of the producers and the planktonic herbivores — is difficult, though something might be achieved using the techniques mentioned on p. 236. As far as larger consumers are concerned, this separation is purely an academic consideration, as they will presumably eat any plankton indiscriminately. Using a well established and balanced aquarium as a model, one might be able to take this study further by collecting some information about the productivity (P_3) and the respiration rate (R_3) of the secondary consumers.

Land ecosystems may be studied more directly. Estimation of standing crop of the producers can generally be made without much difficulty, but careful thought is needed to devise suitable methods of assessing how much this has increased in a known time. Only with annuals, germinating in spring, can one assume that the standing crop in summer represents the season's growth to date (what has not been eaten of P_1, i.e. NU_1 in Fig. 12.4). Many species, such as grasses, are winter-green and make some new growth whenever winter conditions are mild enough, before the real growing season. In other cases, where growth of the shoot starts anew each spring, the first flush is subsidized from underground storage organs, representing part of the previous year's production. Later in the season there may be little further growth above ground, as excess production is all being directed below, forming storage for next year's spring flush. This is particularly evident in some woodland herbs such as Solomon's seal (*Polygonatum multiflorum*). In this species the new season's shoot grows up within a very short time, and the leaves are unfolded before the buds of the trees open. Once the leaves are expanded, however, no further increase in photosynthetic surface takes place throughout the long season in which the plant is in leaf. Production in excess of respiration requirements (P_1) is all stored in the rhizome, and each year the store becomes a little larger, enabling the plant to expose a larger surface for photosynthesis the following year. This continues until the plant is fully grown, when flowering, fruiting and the formation of new branch rhizomes account for excess production. Clearly, in cases like this measurement of the rapid shoot growth in spring would give a very misleading picture of the plant's productivity, which is in fact very low. (A graph of the total leaf area of a Solomon's seal plant in successive years resembles that for a sunflower in which weeks, instead of years, are taken as the units of the time axis.) Productivity could best be assessed on an annual basis, e.g. from the weights of rhizomes. For deciduous woody plants, one might regard the current season's new crop of leaves as an expression of the excess production in the previous season, which has been stored over the winter. So, production in 1967 = biomass of 1968 new leaf crop + increase in dry-weight of stem and roots during 1967. This would of course not apply where plants grow further leaves during the summer season.

Grassland probably lends itself best to studies of terrestrial productivity, for we can take periodic harvests by clipping small sample areas (or even use experimental boxes), and so get a direct measure of further growth of shoots at intervals during the season. Being adapted to withstand grazing the grass remains vigorous under such treatment, where other kinds of plant would be severely damaged. Some allowance must be made for the growth of underground portions, e.g. by estimating the biomass of live roots washed out from a known volume of soil each time the shoots are harvested, as Golley did. From biomass expressed as dry-weight the results can be converted to energy measurements if we determine the calorific value of the shoot and root material. This gives an estimate of that part of the net productivity (P_1) which is not being used by the herbivores (i.e. NU_1 in Fig. 12.4). Respiration of unit area of grassland could be estimated by measuring the rate of production of carbon dioxide, perhaps more conveniently if grassland samples established in flower pots are used. The classic Pettenkofer apparatus could be used, in which the carbon dioxide is absorbed in barium hydroxide solution, and estimated by titration against standard hydrochloric acid, using phenolphthalein as indicator (Baron, 1967).

If we are to estimate the net productivity (P_1) from which, by adding R_1, the gross productivity (P_g) may be derived, we need to know the rate at which available plant material is being consumed by the herbivores. This involves collecting data on the animal populations and their feeding behaviour, a task which soon assumes the proportions of a full-time research project if we are to expect reliable results. It is, however, worth trying out on a small scale the simpler techniques for estimating populations of invertebrates in the grass, more to gain some insight into principles than for the quantitative data themselves. Taken together with estimates of respiration rates made from individuals or small samples of invertebrates, these results will give some rough idea of the scale on which consumer respiration is going on. Similarly, measurements of carbon dioxide production from litter and topsoil can be made to illustrate respiration of the decomposer community. Details of methods for estimating population size, such as removal sampling, mark-and-recapture techniques etc., are given in Odum (1963) and Dowdeswell (1959); the latter also describes techniques for measuring respiration rates of small animals, while the Nuffield Foundation Project's O-level Biology Text and Teachers' Guide (Year III, 1966) gives further details.

Another approach is to make a detailed study of the energy flow in a simple model situation, leaving aside all the complexities and unknowns of a natural ecosystem, yet illustrating the same fundamental principles. Locusts eating grass would serve as a model, but plants with larger leaves are more convenient for productivity measurements, so caterpillars feeding on some suitable host plant might be a better choice.

Primary productivity can be assessed using a modified version of 'Sachs' half-leaf method. The basic technique involves cutting a set of discs (about 15–20 mm. diameter) from one side of the midrib of a leaf, using a suitable cork-borer; leaving the plant to photosynthesize for a known time (5–6 hours), and then collecting a corresponding set of discs from the other side of the leaf. The dry-weights of different sets of discs are then compared (*a*) for plants left in normal daylight and (*b*) plants kept in darkness during the 5–6 hours between taking the first and second set of discs. Leaves at a corresponding stage of development should be chosen for the comparison. Each set of discs represents exactly the same leaf area, and this figure must be computed. Discrepancies from changes in cell turgor during the experiment are hardly likely to introduce an important source of error. The gain in dry-weight of discs from plants kept in daylight represents the net production (P_1) of carbohydrate by known leaf area in a known time. The carbohydrate usage in respiration (R_1) by this same area of leaf tissue is given by the loss in dry-weight of discs taken from plants which were kept in darkness. This figure will in fact be too high, for it will include also carbohydrate translocated away from the leaf during the experiment, but allowance can be made for this if required. By cutting some of the side veins, and thus interfering with translocation in part of the leaf, the rate of removal of carbohydrate from unit area can be estimated.

The net productivity (P_1) added to the respiration rate (R_1), both expressed in terms of gain or loss in dry-weight per unit area of leaf per hour, gives an estimate of the gross productivity (P_g). From this we can compute the daily net productivity, as the outcome of gross productivity (P_g) for the number of hours daylight at the time, less respiration (R_1) for 24 hours. This is assuming that these processes proceed at an even pace throughout the day. From determinations of the calorific value of leaf tissue, details in Nuffield Foundation Project's O-level Biology, Year III (1966), or by using the theoretical value in the carbohydrate equation (p. 238) these rates can be expressed in terms of Kilo-calories per unit area of leaf per hour. It would be interesting to compare the gross productivity so obtained with published data on the amount of solar energy reaching unit area per hour at the appropriate season of the year, and so get a first approximation of the efficiency of the food plant in fixing solar energy. Finally, an estimate of the total leaf area will allow us to work out an approximate figure for the performance of the experimental plant as a whole. A useful device for estimating the area of leaves on the growing plant is a thin metal sheet with a grid of holes ($\frac{1}{16}$ in. is a suitable size) drilled at $\frac{1}{2}$ cm. intervals. This sheet is held parallel to the leaf surface and a few inches from it, directly facing the sun, and a count made of the number of sunflecks on the leaf. If one thinks of each hole in the plate as the centre of a square of side $\frac{1}{2}$ cm. it is clear that:

$$\text{(Number of sunflecks counted)} \div 4$$
$$= \text{leaf area in square centimetres.}$$

We have now obtained figures for the net and gross productivity of the food plant, and we know its energy requirements for respiration. It should not be very difficult to estimate the leaf area consumed daily by the caterpillars and deduce their energy intake from conversion figures already available. It remains to find what use the caterpillars make of this energy. A part is never really digested and made available to the caterpillars (NA_2); this is evident as the potential energy of the droppings, and we can assess it from the dry-weight of droppings produced daily, and their calorific value. Respiration rates per unit fresh weight of caterpillar can be determined using respirometers described in the Nuffield Foundation Project's O-level Biology Text and Teachers' Guide, Year III (1966). If we assume that carbohydrate metabolism still predominates, the rate of carbon dioxide production measured in these techniques can be converted to rate of energy release (R_2). A refinement would be to check the respiratory quotient of the caterpillars, and adjust conversion values if this departs much from unity. That part of the energy not yet accounted for should appear as new caterpillar tissue (or shed skins). By weighing the caterpillars at intervals we can obtain growth curves for them. A few must be sacrificed to derive conversion figures relating fresh-weight to dry-weight, and to determine the calorific value of caterpillar tissue: the growth curves can then be expressed in terms of gain in fixed energy per unit dry-weight of caterpillar per day (P_2).

We have here the means to build up from first-hand observations

I_2

a balance sheet accounting for energy flow through this simple producer-consumer system. The situation is artificial in its simplicity; the effects of so many interacting populations of predators, rival herbivores and competing plants, which would be present in the natural ecosystem, are left out. Yet, under natural conditions, the same caterpillars would still eat the same food plant, and the immediate fate of their energy income would be much the same. The simplicity of the model, and our first-hand practical involvement in studying it, can give an insight into this basic idea of energy flow which helps greatly in our understanding of the complexities of natural ecosystems.

APPENDIX

SAMPLING SOIL ATMOSPHERE

This technique was devised by J. L. Harley and J. K. Brierley (1953), to whom the author is indebted for permission to reproduce Fig. 1, and for the following description which is taken directly from the account published in the *Journal of Ecology*.

The apparatus is shown in Fig. 1. The gas-sampling tube consists of a length of polyvinyl chloride tubing* A, about 30 cm. long and 3 cm. bore, which gives two 25 c.c. gas samples for analysis. In one end is placed a rubber bung B, perforated by a capillary tube C, 1 mm. in bore. On this a stopper D, made from a length of rubber tubing and a piece of glass rod, could be fixed. The rubber bung E, is inserted in the other end of the tube, and through this a hole is bored to carry a piece of glass tubing F, $1\frac{1}{2}$ cm. in diameter. On F is fitted about 6 cm. of rubber tubing G; this is called the nozzle. To open and close the nozzle a well-greased screw-clip H, is used. Finally a piece of strong wire I, is soldered to the clamping screw of the clip, thus enabling the nozzle to be opened and closed when the sampling tube is in position in the soil. The bottom bar of the clip should be fixed, in order to prevent the clip from twisting open when closing the nozzle. When completely sealed with the nozzle closed and D in place, the tube can be tested for leaks by squeezing.

Harley and Brierley used sampling tubes in pairs; one tube of each pair was filled with nitrogen and the other with air. This provides a check that they are left in the soil long enough to obtain a true sample of soil atmosphere. The tubes were placed in the soil side-by-side with the rubber nozzles closed, and the wire I, projecting above the soil, was appropriately labelled. After allowing time for the soil atmosphere to regain equilibrium following the disturbance of burying the tubes, they were opened by carefully twisting I, and thus opening the rubber nozzle. The tubes

* Polyvinyl chloride tubing obtained from BX Plastics Ltd., Higham Station Avenue, Chingford, London, E.4.

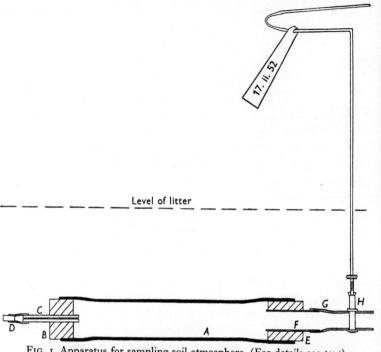

FIG. 1. Apparatus for sampling soil atmosphere. (For details see text)

From Harley and Brierly: A Method of Estimation of Oxygen and Carbon Dioxide Concen-
trations in the Letter Layer of Beech Woods: Journal of Ecology, 41

were left in the soil for at least a month, then sealed by closing the nozzle, and removed.

The advantage of the plastic sampling tube is that the gas sample can easily be squeezed into a gas analysis apparatus through the capillary tube C (after first removing the stopper D). Ideally, a Barcroft-Haldane apparatus would be used for gas analysis, as described by Peters and Van Slyke (1932), and figured in M. G. Brown and W. H. Dowdeswell (1956); but simpler, if less accurate techniques for the absorption of oxygen and carbon dioxide are available (Nuffield O-level Course, Year III).

If comparable results are obtained from both sample tubes of each pair used, one can feel assured that the tubes have been in place long enough to reach equilibrium with the soil atmosphere, as they approach it from widely different starting points. It seems likely that in most soils equilibrium is reached in a sampling time of much less than a month.

MEASUREMENT OF SOIL TEMPERATURE

Soil temperature is most conveniently measured using a therm-istor*. This device consists of a small capsule containing some compound of which the electrical resistance falls rapidly with increasing temperature. All that is necessary, then, is to measure the resistance of the thermistor accurately by means of a Wheat-stone Bridge circuit (Fig. 2). R_1 and R_2 are suitably balanced known resistances. The variable resistance R_3 is adjusted until no

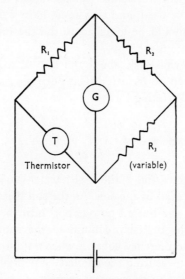

FIG. 2. Wheatstone bridge circuit for use with thermistor

current flows through the galvanometer G, and the resistance of the thermistor T is then given by $T = R_3 \times R_1 / R_2$. A Post Office box provides all the necessary resistances in a conveniently port-able form, and an ordinary torch battery will serve perfectly well as the source of current. Only a moving-coil galvanometer is needed to complete the outfit. The entire apparatus may now be obtained as a single compact unit in the form of a 'Logohm' resistance meter, from which resistances are read off directly.

To convert the resistances into temperature readings the therm-

* Made by Standard Telephones and Cables Ltd., Footscray, Sidcup, Kent.

istor must be calibrated. Its resistance when immersed in a water bath at different known temperatures (measured by a sensitive thermometer) is determined carefully, and from these results a conversion graph can be drawn.

Thermistors have the great advantages of being small, inexpensive and not subject to damage by water. They can be left buried in the soil with only the leads projecting, or let down into the water of a pond or canal to determine the temperature at different depths. Low resistance leads soldered to the thermistor should of course be used, and they must have waterproof insulation. The leads should be attached to the thermistor when it is calibrated.

Tiny 'bead thermistors', about the size of a pin's head, are obtainable, and can be used for such purposes as measuring the temperature inside resting buds.

MEASUREMENT OF CAPILLARY POTENTIAL OF THE SOIL
(pF)

A soil tensiometer (Fig. 3) filled with water is used to determine the capillary potential of the soil. As the soil dries it will tend to draw water from the buried porous pot until the 'suction force' is balanced by the mercury column in the capillary tube EF. The height of the mercury column when equilibrium is reached gives a means of measuring the capillary potential of the soil, that is, the force needed to draw water from it. It is obviously more convenient to use a mercury column in the soil tensiometer, though actually capillary potential is defined as the height of the *water* column (in cm.) needed to draw water from the soil. This is obtained by multiplying the height of the mercury column (in cm.) by the factor 13·6 (relative density of mercury). Capillary potential is often expressed as the pF of the soil, or the *logarithm* of the height of the water column (in cm.). Two tensiometers, with porous pots buried at different depths can give information about water movement in the soil, for example, the penetration of rain into a dry soil, or the region being exploited by roots.

In the apparatus shown in Fig. 3, AB and CD are of ordinary glass tubing, with pressure tubing joints at A and C. CEF is of capillary tubing, about 1 mm. bore: EF should be about seventy-

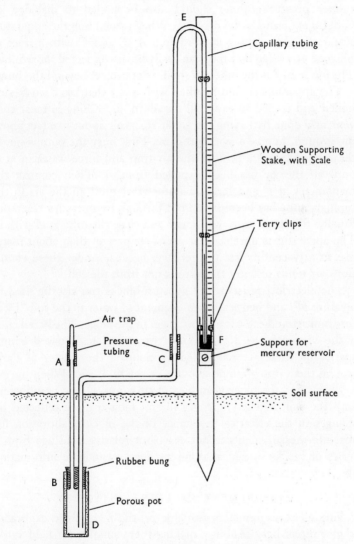

FIG. 3. Soil tensiometer for measurement of capillary potential (pF)
of soil

six centimetres long. The porous pot* should be boiled to expel any air trapped in the pores and allowed to cool in water. New rubber pressure tubing should also be boiled to dislodge air bubbles adhering to the dressing. When assembling the apparatus, smear the glass tubing at the joints A, B and C with a trace of Silicone grease, so that they can easily be disconnected for refilling. The lower end of the tube AB should not project beyond the bung.

The apparatus should be filled with water that has been freshly boiled and cooled to expel all dissolved air. Filling is most conveniently done by having the water reservoir about five feet above the ground, to give a small pressure. First bury the porous pot to the required depth, remove the air trap and force water in at C until all the air has been expelled from A. Then connect the manometer at C and force in water at A until all the air in the capillary tube has been driven out through the mercury reservoir. Finally, fill the air trap with water and close the tube at A with it. The apparatus is unreliable at pressures greater than about forty-five to fifty centimetres of mercury, because under these conditions air tends to enter the porous pot from the soil.

The electrical resistance of gypsum blocks can also be used to measure pF, and hence study movement of water in the soil. Twin core multistrand cable is used as a lead to the block. The bared ends of the wires are teased out to form two flat 'plates', spaced 1 mm. apart, and these are embedded by pouring the plaster of Paris around them into a mould about 1 in. $\times \frac{1}{2}$ in. $\times \frac{1}{3}$ in. When buried in the ground the block absorbs moisture until it is in equilibrium with the soil, and any changes in soil moisture are reflected by changes in the electrical resistance of the block. Calibration for absolute measurements is not easy, but relative readings from a series of blocks buried at different depths can give information about changes in soil moisture.

MEASUREMENT OF LIGHT INTENSITY

Various exposuremeters available on the market are calibrated to give direct light readings, making them suitable for field work.

Photo-voltaic cells can also be obtained from several manufacturers† for the price of a few shillings, and, in conjunction with

---

* Obtainable from Doulton and Co., Albert Embankment, London, S.E. 11. Porcelain of Grade P 13 A is suitable.

† For example, Messrs. Megatron Ltd., 115A Fonthill Road, London, N. 14.

a sensitive moving coil milliammeter one of these can easily be used to build a reliable light meter. Full details of the technique of assembling, and mounting the cell in a 'Perspex' frame are given by Dowdeswell and Humby (1953). The cell and leads are made waterproof, so that measurements can be taken under water if needed, as, for example, in plankton studies, or investigations of submerged aquatics. Having the cell separate from the milliammeter is also an advantage in taking readings under dense vegetation, where it might be impossible to see the dial of an ordinary exposuremeter.

The range of the instrument can be increased by mounting the milliammeter in a small box with shunts which can be put into the circuit for measuring higher light intensities. This is more reliable than reducing the surface of the cell exposed to the light, because the sensitivity of different parts often varies.

When studying shade cast, and plant competition, as in woodlands, light measurements are usually expressed as percentages of full light in the open, at the time. This raises the problem of how best to integrate the mosaic of sunflecks and shadows of varying intensities which make up the light-climate of the woodland floor. The simplest way is to take the mean of a large number of readings taken over the area being studied. G. C. Evans (1956) has proposed a method whereby the cell is held face downwards at a fixed distance above a whitened board of standard size. This board acts as a reflector, and at the same time serves to integrate local variations in light intensity.

MEASUREMENT OF THE EVAPORATING POWER OF THE AIR

This can be measured in absolute terms as **saturation deficit**, derived by using tables from the readings of a wet and dry bulb thermometer. A convenient *comparative* measure is given by using an atmometer, which assesses directly the rate of evaporation from an exposed surface of wet porous pot. As the surface area of the porous pot is not a known standard quantity, the rates of evaporation from the *same* atmometer are usually compared under different conditions, for example, two different habitats.

There are many different forms of atmometer: the two shown in Fig. 4 were designed by Prof. W. O. James of the Department of Botany, Imperial College of Science and Technology, London.

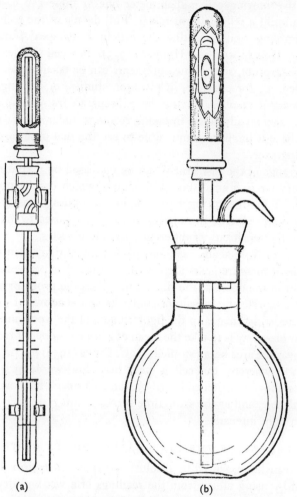

(a)   (b)

FIG. 4. (a) Atmometer for rapid determinations of water loss, useful in comparing different microhabitats. (b) Atmometer for determining the weight of water lost over considerable periods

From A. G. Tansley: Introduction to Plant Ecology: Allen and Unwin

For rapid readings type *A* is preferable. It consists essentially of a porous 'candle', connected to a length of one millimetre bore capillary tubing, with a bubble trap at the top. The whole is filled with *distilled* water, and this is drawn up from the reservoir to replace evaporation losses from the porous surface of the candle.

When the reservoir is lowered momentarily, an air bubble is introduced into the capillary tube: its rate of progress up the tube can be measured against the scale fixed to the frame of the instrument, and is proportional to the rate of evaporation from the porous candle. As there is often considerable fluctuation due to gusts of wind, the average of several readings should be taken.

Type *B* has a much larger reservoir and is used to record loss in weight due to evaporation over longer periods. The device at the top of the tube is a valve (working on the same principle as that of a bicycle tyre) to prevent water from getting back into the apparatus when it is exposed to rain.

COMPARISON OF TRANSPIRATION RATES USING COBALT CHLORIDE PAPER

This technique depends upon the fact that anhydrous cobalt chloride is a deep blue colour, but it changes to pale pink as the salt takes up its water of crystallization from the air. Filter paper which has been soaked in a solution of cobalt chloride and allowed to dry assumes this deep blue colour when it is dried by a small flame. If it is then held pressed against the surface of a leaf between two glass slides, gripped by a small bulldog clip (Fig. 5*a*), the time taken for the paper to turn pink will depend upon how fast the leaf is supplying water vapour through the stomata. Using similar samples of cobalt chloride paper the method allows a rough comparison of the rates of transpiration from upper and lower surfaces and between different leaves.

Accuracy can be considerably improved by setting the cobalt chloride paper between colour standards, and timing the change from one colour standard to the other. This requires good matching of the standards, and the following notes, taken from a paper by Henderson (1936) give some account of the preparation of these three-colour strips. A smooth, high-grade filter paper (for example, Whatman's No. 1) is essential for even colouring. For the cobalt chloride paper a solution of 150 gm. per litre is used. Both colour standards are made from a stock solution of methylene blue, containing 1 gm. per litre. For the dark blue standard this solution is diluted to $\frac{1}{8}$ full strength, and for the light blue standard to $\frac{1}{32}$ full strength followed by treatment in a 1 in 10,000 solution of eosin to tint it faintly pink.

Each sheet of filter paper is first soaked in water for one minute, then pressed between sheets of blotting paper with a soft rubber roller (squeegee). It is then immersed for one minute in the appropriate dye, removed, and squeegeed again. The light blue standard is given a second, similar treatment in very dilute eosin solution, as mentioned above. The sheets are now pinned up to dry.

When dry the coloured papers are cut into narrow strips about three millimetres wide and assembled as shown in Fig. 5b. Narrow strips of black Passe Partout, stuck transversely across the parallel serve to hold them together. Finally the sheets are cut up into

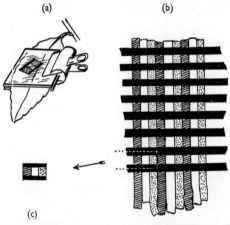

(a)   (b)

(c)

FIG. 5. Three-colour strips with cobalt chloride paper. (a) Method of clipping onto a leaf (which must be supported) between pieces of glass slide. (b) and (c) stages in preparation of the strips. (See text)

individual three-colour strips (Fig. 5c), with the cobalt chloride paper in the middle between the two colour standards.

Before use the three-colour strips are dried beside a small flame until the cobalt chloride paper assumes a *deeper* blue than the dark blue colour standard. They are then put immediately into a specimen tube which has some calcium chloride beneath a piece of wire gauze at the bottom, and the tube is securely corked. In this way they can easily be transported and used in the field. It should be emphasized that the strips should always be picked up with forceps, and can only be used on dry leaves, as any free moisture will dissolve out the cobalt chloride, ruining both the results and the three-colour strip. With careful treatment the strips can be used again and again.

Where different leaves transpiring from both surfaces are to be compared, the *reciprocals* of the times taken for the colour change on the upper and lower surfaces should be added as a basis of comparison. To take a hypothetical example:

Plant investigated	Time taken for cobalt chloride paper to change from dark blue to light blue standard	
	Upper surface of leaf	Lower surface of leaf
Species *A*	35 minutes	5 minutes
Species *B*	12 minutes	8 minutes

In one minute, species *A* loses from unit area of the upper surface of the leaf $\frac{1}{35}$ of the 'dose' of water needed to change the colour of the cobalt chloride paper from one standard to the other. From unit area of the lower surface it loses $\frac{1}{5}$ of a 'dose'— the total loss of water from both surfaces is then $\frac{1}{35}+\frac{1}{5}=\frac{8}{35}$ or 0·23 'dose'. In comparison the total loss from species *B* is $\frac{1}{12}+\frac{1}{8}=\frac{5}{24}$ or 0·21 'dose', that is, species *A* is transpiring a little more rapidly than species *B* per unit area of leaf surface.

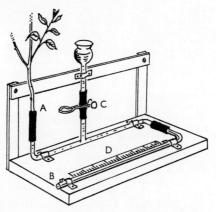

FIG. 6. Simple potometer suitable for class use

In working with cobalt chloride papers one must bear in mind that the technique eliminates any differences in microclimate such as between an exposed, windy habitat and one that is sheltered and moist — both are replaced by a common external environment of dry filter paper pressed against the leaf. Results collected from two habitats giving widely different atmometer readings may therefore be misleading.

POTOMETERS

The principle of the potometer is so well known that no account of its use is needed here. It does, however, seem worth while commenting briefly on two of the many designs that are used. The

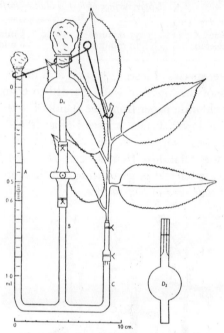

FIG. 7. Potometer allowing simultaneous measurement of both transpiration and water uptake

From G. C. Evans: Transpiration and Water Uptake of Cut Shoots: Jnl. of Ecology, 37, 1

type shown in Fig. 6 is easily made in the school workshop and proves very satisfactory for class use. One of its advantages that is not always appreciated is that the place of insertion of the cut shoot at *A* is at a higher level than the end of the capillary tube *B*. Any slight leak at *A* will therefore result in water movement in the *opposite* direction from that indicating water uptake by the shoot. When the apparatus is assembled, the meniscus can be brought to the middle of the scale D by blowing down B, and at the same time momentarily releasing the clip C. If the meniscus then moves towards B there is a leak; if away from B the apparatus is working properly.

It will of course be realized that potometers do not actually measure transpiration, but water uptake by a cut shoot, though the discrepancy between these two values is usually very small. An apparatus has been described by Evans (1949) by which both transpiration rate (as loss in weight) and water uptake (by volume) can be measured simultaneously, thus giving data about the state of turgor of the tissues. This is shown in Fig. 7.

ECOLOGICAL HERBARIUM

An attempt to build up a systematic herbarium is so massive a task that it is likely to meet with discouragement in a school before much has been achieved; but collection of characteristic species from well-defined associations in the locality are well within the scope, and are in many ways more useful. In starting, it would be wise to concentrate on a single association. Only the *dominant* or *commoner species* and *indicator species* need be collected; the rarer one should be omitted deliberately as making little contribution to the community. Mosses forming prominent societies may well be included with the flowering plants, as certain species are coming to be regarded as reliable indicators of habitat factors. For identification of mosses reference can be made to E. V. Watson (1957). The little booklet on woodland mosses brought out by the Forestry Commission is also useful (H. Watson, 1947).

When collecting flowering plants; roots or rhizomes should be gathered as well as the aerial parts; fruits as well as flowers, if possible. Careful notes should be made of the exact locality, and the status of the plant in the association — to which layer it belonged, if in a stratified community, whether forming a local society, relative abundance, and any factors which appear to determine its local occurrence, etc. This data should later be incorporated on the herbarium sheet label. Collections should not be arranged systematically, but according to the associations from which they are drawn. The same species, if it is widespread, may thus appear repeatedly in collections from different associations. This brings us to a second purpose that an ecological herbarium can serve — as a record of ecotype variation. This will involve specialized collections of transplanted material grown under different conditions (p. 139), but the differences seen in specimens from various associations in the general ecological herbarium may well provide extra information.

The technique of pressing is straightforward. The plants are laid out carefully on newspaper, in as natural a position as possible (any very thick or juicy parts should first be sliced in half), and the specimens are then pressed between several layers of newspaper. This should be changed every two or three days, until the specimens are really dry. The actual press consists of a pair of boards (five-ply is suitable) the size of the herbarium sheets to be used ($10\frac{1}{2} \times 16\frac{1}{2}$ in. is a common size), and held together by stout webbing straps. These may be bored with a few large holes to help allow the moisture to escape, or a strong trellis of slats set at right angles may be used instead of boards, but provided the paper is changed frequently, the simple boards give quite good results.

When properly dry, the specimens are mounted on herbarium sheets of cartridge paper. The easiest way to fix them is by strips of gummed paper or 'Cellotape'. It is important that all the notes about habitat, status, etc., should appear on the herbarium sheet, and some system of numbering is useful to ensure that specimens are not confused during pressing.

The herbarium sheets should be kept in a dry place, with moth balls of para-dichlorobenzine ('P.D.B.' of chemist's shops) to protect from insect attack. If mites or insects are found the specimens should be painted with a solution of mercuric chloride in alcohol.

Mosses do not require pressing: they are best just left to dry, and then stored in suitable envelopes or folded slips of paper. They can be soaked out in water later for examination.

ESTIMATION OF COMPENSATION POINT
USING 'NUFFIELD' BICARBONATE INDICATOR

This extremely sensitive indicator is described in Year III Teacher's Guide of the Nuffield Foundation O-level Biology Course (1966). At pH 8·2, in equilibrium with the 0·03 per cent CO_2 in the air, the indicator is red. It turns purple at pH 8·4, when CO_2 is removed by excess photosynthesis; and with the addition of CO_2 from excess respiration it turns through orange (pH 7·8) to yellow (pH 7·2). We have in it a simple means of following these changes in a variety of situations (e.g. in sunny and shaded parts of a pond throughout the day).

To estimate the compensation point* of a plant, small whole leaves, or leaf discs, are enclosed with about 5 c.c. of indicator in a number of stoppered test-tubes. The leaves of land plants must, of course, be supported above the surface of the indicator by means of a loose cotton wool plug, but submerged water plants are best placed *in* the indicator solution. Tubes and stoppers should be rinsed with distilled water and then with indicator solution before setting up, and the cotton wool plugs and leaf discs should be inserted using clean forceps. The tubes are arranged in a series at different distances from a bright lamp, and the light intensity determined at which the colour of the indicator remains unchanged. No colour change, means that the CO_2 output (respiration) is balanced by CO_2 uptake (photo-synthesis), i.e. the leaf is at compensation point.

Some plants give variable results, possibly due to partial closure of the stomata, for submerged aquatics give more consistent values.

WINKLER METHOD FOR DETERMINATION OF OXYGEN DISSOLVED IN WATER

Apparatus needed in the field:

A 20 c.c. disposable plastic syringe, with a glass bead inside to aid mixing; and a long needle.

Storage bottles and a thermometer.

Reagents:
  (i) 40 per cent manganous chloride solution.
  (ii) alkaline iodide solution: 32 gm. sodium hydroxide plus 10 gm. potassium iodide made up to 100 c.c. with distilled water.
  (iii) concentrated orthophosphoric acid.

Procedure:

1. Record the temperature of the water being tested.
2. Fill the dead space in the syringe with manganous chloride solution; expel any air bubbles.

---

* This application of the technique used in the Nuffield O-level Biology Course was developed by Mr. A. K. Rawlins.

3. Draw up into the syringe 18 c.c. of the water sample, taking care to avoid the inclusion of any air bubbles.
4. Draw up about 0·5 c.c. of alkaline iodide solution; rock the syringe to mix thoroughly, and leave for at least 5 minutes to allow absorption of the dissolved oxygen by the precipitate.
5. Draw up enough phosphoric acid (about 0·5 c.c.) to dissolve the precipitate, giving a clear golden brown liquid.
6. The oxygen is now 'fixed', and an equivalent quantity of iodine has been released. Transfer all the solution from the syringe to a stoppered bottle and retain for titration in the laboratory.
7. For titration, take the entire sample from the bottle (washing out with distilled water into the titration flask) and titrate with N/800 sodium thiosulphate (0·3102 gm. per litre) until the yellow colour of the iodine has almost gone. Then add a few drops of freshly made starch solution as an indicator, and titrate till the colour disappears.

$$\frac{\text{Volume of thiosulphate (c.c.)} \times 10}{\text{Volume of water sample (18 c.c.)}} = \begin{array}{c}\text{mg. oxygen per litre of}\\ \text{water sample.}\end{array}$$

(To convert mg. oxygen per litre to c.c. per litre (at N.T.P.), multiply by $\frac{22\cdot4}{32}$.)

BIBLIOGRAPHY

ADAMSON, R. S., 'The Woodlands of Ditcham Park, Hampshire,' *J. Ecol.*, **9**, 114–219 (1921).

ANDERSON, V. L., 'The Water Economy of Chalk Plants', *J. Ecol.*, **15**, 72–129 (1927).

ARBER, A., 'Water Plants' (Cambridge University Press, 1920).

ASHBEL, D., 'Solar Radiation and Soil Temperature' (Jerusalem, 1942), quoted in *J. Ecol.*, **35**, 142 (1947).

ASHBY, E., 'Modern Concepts of Xerophytes', *School Sci. Rev.*, No. 55, 329–344 (1933).

ASHBY, E., 'Experimental Work on Xeromorphic Characters', *School Sci. Rev.*, No. 60, 509–512 (1934).

BARON, W. M. M., 'Organization in Plants' (London, Arnold, 2nd Ed. 1967).

BATES, G. H., 'The Vegetation of Footpaths, Side walks, Cart-tracks and Gateways', *J. Ecol.*, **23**, 470–487 (1935).

BILLINGS, W. D., 'Plants and the Ecosystem' (London, Macmillan, 1964).

BLACKMAN, G. E. and RUTTER, A. J., 'Physiological and Ecological Studies in the Analysis of Plant Environment', *Ann. Bot. (London) New Series*, **10**, 361–390 (1946); **11**, 126–158 (1947); **12**, 1–26 (1948); **13**, 453–489 (1949); **14**, 487–520 (1950).

BLACKMAN, G. E. and RUTTER, A. J., 'Biological Flora of the British Isles—*Endymion nonscriptus*', *J. Ecol.*, **42**, 2, 629–638 (1956).

Blundell's School Sci. Mag., **2**, 29–33 (1947), 'The Flora and Fauna of Rain-Gutters' by G. O. Mackie.

5, 49–52 (1950), 'An Experiment to find the Effect on the Growth of an Annual Plant when Flowering is Prevented' by P. R. Ash.

6, 55–59 (1951), 'Observations on the Flora of some Walls near the School' by I. G. Turner.

6, 60–63 (1951), 'An Examination of the Factors Influencing the Earthworm Population in the Soil' by M. Jones.

7, 26–41 (1951–2), 'Studies on the Biology of Ivy-leaved Duckweed (*Lemna trisulca*)' by M. J. Corrigan.

7, 42–46 (1951–2), 'The Pollination of Orchids' by D. A. P. Butcher.

8, 27–32 (1952–3), 'Observations on the Circulation of Nitrogen in the Grand Western Canal' by C. L. Honeybourne.

8, 32–36 (1952–3), 'Soil Temperatures and Plant Growth'.

9, 25–36 (1953–4), 'Physical Factors in Seed Dispersal' by R. W. Moore.

10, 20–26 (1954–5), 'An Investigation of the Factors influencing the Luxuriance of the Field Layer on Different Slopes in Eppscleave Wood, Devon' by D. P. T. Burke.

Blundell's School Sci. Mag., **10**, 26–30 (1954–5), 'An Investigation of some of the Changes taking place in the Rotting of a Grass Compost Heap' by R. Hull.

12, 21–34 (1956–7), 'An Investigation of the Factors Underlying the Difference in Vegetation on the North- and South-facing Slopes of a Steep Valley in Spring' by R. G. Pembrey.

12, 34–38 (1956–7), 'An Investigation of the Function of Bud-Scales in Flowering Currant by S. M. Vaughan.

BRACHER, R., 'Ecology in Town and Classroom' (Bristol, Arrowsmith. 1937).

BRENCHLEY, W. E., 'Buried Weed Seeds' *J. Agric. Sci.*, **9**, No. 1 (1918),

BRIERLEY, J. K., 'Some Preliminary Observations on the Ecology of Pit Heaps', *J. Ecol.*, **44**, 2, 383–390 (1956).

BROWN, M. G. and DOWDESWELL, W. H., 'Apparatus for the Ecological Study of Soil and Mud', *School Sci. Rev.*, No. 134, 70–77 (1956).

CHAMBERS, E. G., 'Statistical Calculation for Beginners' (Cambridge University Press, 1955).

CHIPPINDALE, H. G., 'The Effects of the Floods of 1953 on the Agriculture of Eastern England' (*New Biology*, No. 24, 33–49, 1957).

CONRAD, J. P. and VEIHMEYER, F. H., 'Root Development and Soil Moistures', *Hilgardia*, **4**, 113–134 (1929).

DOWDESWELL, W. H., 'Practical Animal Ecology' (London, Methuen, 1959 and 1968).

DOWDESWELL, W. H. and HUMBY, S. R., 'A Photo-voltaic Light Meter for School Use', *School Sci. Rev.*, No. 125, 64–70 (1953)

DUDDINGTON, C. L., 'The Friendly Fungi' (London, Faber, 1957).

ELTON, C., 'The Ecology of Invasions by Animals and Plants' (London, Methuen, 1958).

EVANS, G. C., 'Note on an Apparatus for Simultaneous Measurement of Transpiration and Water Uptake of Cut Shoots', *J. Ecol.* **37** ; 1, 171–173 (1949).

EVANS, G. C., 'An Area Survey Method of Investigating the Distribution of Light Intensity in Woodlands, with Particular Reference to Sunflecks', *J. Ecol.*, **44**, 2, 391–428 (1956).

FISHER, R. A. and YATES, F., 'Statistical Tables for Biological, Agricultural and Medical Research' (Edinburgh, Oliver and Boyd, 1943).

FITTER, R. S. R., 'London's Natural History' (London, Collins, 1945).

FRITSCH, F. E. and SALISBURY, E. J., 'Plant Form and Function' (London, Bell, 1946).

GILLHAM, MARY, E., 'Ecology of the Pembrokeshire Islands', *J. Ecol.*, **41**, 1, 84–99 (1953); **42**, 2, 296–327 (1954); **43**, 1, 172–206 (1955), **44**, 1, 51–82 (1956); **44**, 2, 429–454 (1956).

GILLHAM, MARY, E., 'Vegetation of the Exe Estuary in Relation to Water Salinity', *J. Ecol.*, **45**, 3, 735–756 (1957).

GIMINGHAM, C. H., 'The effects of Grazing on the Balance between *Erica cinerea* and *Calluna vulgaris* in Upland Heath, and their Morphological Responses', *J. Ecol.*, **37**, 1, 100–119 (1949).

GODWIN, H., 'Dispersal of Pond Floras', *J. Ecol.*, **11**, (1923).

GODWIN, H., 'Pollen Analysis (Palynology)', *Endeavour*, **10**, No. 37 5–16 (1951).

GOLLEY, F. B., 'Energy Dynamics of a Food Chain of an Old-Field Community', Ecological Society of America, *Ecological Monographs*, **30**, No. 2 (1960).

GREIG-SMITH, P., 'Quantitative Plant Ecology' (London, Butterworth, 1957).

HANSON, W. C. and KORNBERG, H. A., 'Radioactivity in Terrestrial Animals near an Atomic Energy Site', Proc. Int. Conf. Peaceful Uses of Atomic Energy, **13**, 385–388 (1956).

HARLEY, J. L. and BRIERLEY, J. K., 'A Method of Estimating Oxygen and Carbon Dioxide Concentrations in the Litter Layer of Beech Woods', *J. Ecol.*, **41**, 2, 385–387 (1953).

HARPER, J. L., 'Approaches to the Study of Plant Competition' Symposium of Society for Experimental Biology No. xv (1961).

HENDERSON, F. Y., 'The Preparation of "Three-colour" Strips for Transpiration Measurement', *Ann. Bot.* (*London*), **50**, No. 198, 321–324 (1936).

HEPBURN, I., 'Flowers of the Coast' (London, Collins, 1952).

HESLOP-HARRISON, J., 'New Concepts in Flowering-Plant Taxonomy, (London, Heinemann, 1953).

JACKSON, G. and SHELDON, J., 'The Vegetation of the Magnesian Limestone Cliffs at Markland Grips near Sheffield', *J. Ecol.* **37**, 1, 38–50 (1949).

KERSHAW, A. K., 'Quantitative and Dynamic Ecology' (London, Arnold, 1964).

KEVAN, D. K. MCE., 'Soil Animals' (London, Witherby, 1962)

KLEIBER, M., 'The Fire of Life' (New York, Wiley, 1961).

LACK, D., 'Darwin's Finches', *Scientific American*, April, 1953.

LEOPOLD, A., 'A Sand County Almanac' (New York, 1949), quoted in WOODBURY (1954).

MATHER, K., 'Statistical Analysis in Biology' (London, Methuen, 1951).

NUFFIELD SCIENCE TEACHING PROJECT, O-Level Biology Course, Year III, (London, Longmans/Penguin Books, 1966).

ODUM, E. P., 'Fundamentals of Ecology' (Philadelphia, Saunders, 1959).

ODUM, E. P., 'Ecology' (New York, Holt, Rinehart and Winston, 1963).

ODUM, H. T., 'Trophic Structure and Productivity of Silver Springs, Florida,' Ecological Society of America, *Ecological Monographs*, **27**, No. 1 (1957).

OOSTING, H. J., 'The Study of Plant Communities', 2nd Ed., (San Francisco, W. H. Freeman, 1956).

OVINGTON, T. D., 'Vegetation and Water Conservation' (New Biology, No. 25, 47–63, 1958).

PEARSALL, W. H., 'Mountains and Moorlands' (London, Collins, 1950).

PERCIVAL, M. S., 'Floral Biology' (Oxford, Pergamon Press, 1965).

PETERS and VAN SLYKE, 'Quantitative Clinical Chemistry Methods' (London, 1932).

PHILLIPSON, J., 'Ecological Energetics' Institute of Biology, Studies in Biology, No. 1 (London, Arnold, 1966).

POEL, L. W., 'Soil Aeration in Relation to *Pteridium aquilinum*', *J. Ecol.* **39**, 1, 182–191 (1951).

POORE, M. E. D., 'The Use of Phytosociological Methods in Ecological Investigations', *J. Ecol.*, **43**, 1, 226–269 (1955); **43**, 2, 606–651 (1955); **44**, 1, 28–50 (1956).

PRIESTLEY, J. H. and SCOTT, L. I., 'An Introduction to Botany', 3rd Ed. (London, Longmans, Green, 1955).

PRIME, C. T., 'Lords and Ladies' (London, Collins, 1960).

RAUNKIAER, C., 'The Life Forms of Plants and Statistical Plant Geography' (Oxford University Press, 1934).

RISHBETH, J., 'The Flora of Cambridge Walls', *J. Ecol.*, **36**, 1, 136–148 (1948).

SALISBURY, E. J., 'Stratification and *p*H of Soil in Relation to Leaching and Plant Succession, with Special Reference to Woodlands', *J. Ecol.*, **9**, 220–240 (1921).

SALISBURY, E. J., 'The Reproductive Capacity of Plants' (London, Bell, 1942).

SALISBURY, E. J., 'The Living Garden' (London, Bell, 1945).

SALISBURY, E. J., 'Downs and Dunes, their Plant Life and its Environment' (London, Bell, 1952).

SEARS, P. B., 'The Biology of the Living Landscape' (London, Allen and Unwin, 1962).

SNEDECOR, G. W., 'Statistical Methods' (Ames, Iowa State College Press, 1946).

STAMP, L. DUDLEY, 'Man and the Land' (London, Collins, 1955).

TANSLEY, A. G., 'The British Isles and their Vegetation' (Cambridge University Press, 1939).

TANSLEY, A. G., 'Britain's Green Mantle' (London, Allen and Unwin, 1949).

VEIHMEYER, F. H., 'Factors Affecting the Irrigation Requirements of Deciduous Orchards', *Hilgardia*, **2**, 125–284 (1927).

WATSON, E. V., 'British Mosses and Liverworts' (Cambridge University Press, 1957).

WATSON, H., 'Woodland Mosses', Forestry Commission Booklet No. 1 (London, H.M. Stationery Office, 1947).

WATT, A. S. 'On the Ecology of British Beechwoods, II. Development and Structure of Beech Communities on the Sussex Downs', *J. Ecol.*, **13** (1925).

WATT, A. S., 'Pattern and Progress in the Plant Community', *J. Ecol.*, **35**, 1, 1–22 (1947).

WATT, A. S. and FRASER, G. K., 'Tree Roots and the Field Layer', *J. Ecol.*, **21**, 404–414 (1933).

WEAVER, J. E. and CLEMENTS, F. E., 'Plant Ecology' (New York, McGraw-Hill, 1929).

WILLIAMS, W. T. and BARBER, D. A., 'The Functional Significance of Aerenchyma in Plants' from Symposium No. xv of the Society for Experimental Biology: Mechanisms in Biological Competition (1961).

WILLIAMS, W. T. and LAMBERT, J. M., 'Multivariate Methods in Plant Ecology', *J. Ecol.*, **47**, 1, 83–101 (1959); **48**, 3, 689–710 (1960).

WILSON, F., 'The Biological Control of Weeds' (New Biology, No. 8, 51–74, 1950).

WOODBURY, A. M., 'Principles of General Ecology' (New York, Blakiston, 1954).

INDEX

Page numbers in heavy type indicate main references in text.

K